An Introduction to Cold and Ultracold Chemistry

An Introduction to Cold and Ultracold Chemistry

Rose Marie O. Mendoza

ARCLER
PRESS

www.arclerpress.com

An Introduction to Cold and Ultracold Chemistry

Rose Marie O. Mendoza

Arcler Press
224 Shoreacres Road
Burlington, ON L7L 2H2
Canada
www.arclerpress.com
Email: orders@arclereducation.com

ISBN: 978-1-77469-135-9 (Hardcover)

ABOUT THE AUTHOR

Dr. Rose obtained her PhD in Chemical Engineering from the University of the Philippines-Diliman in 2013, Her Masters in Chemical Engineering and BS Chemical Engineering degree from Adamson University. She is also a Professor in the Graduate School Department under the Master of Engineering Program at Adamson University since 2006, and is a Visiting Research Fellow at the Department of Environmental Engineering and Science, and the Department of Environmental Resource Management in Chia Nan University of Pharmacy and Science, Tainan Taiwan since 2010. She obtained her Post Doctorate Degree in Green Power: Hydrogen Generation and Fuel Cell Development from the University of California Merced, USA (in collaboration with the Energy Storage and Conversion Materials Laboratory of the University of the Philippines, Diliman) and is presently also a collaborating researcher at the Kindai University, Nara Japan, focusing on applications of electric and electrostatic field.

Dr. Mendoza has several international and local researches covering Fuel cells and solid oxide electrolysis cells (SOECs), green energy and green technology, biomedical and natural products, fuels and fuel development, at water and wastewater treatment, recovery and remediation. She also published several international textbooks on water management, introductory chemical engineering and chemistry. Presently, Dr. Mendoza is the Chief Science Officer of Mark-Energy Revolution Corporation, Philippines.

TABLE OF CONTENTS

LIST OF FIGURES

LIST OF TABLES

LIST OF ABBREVIATIONS

APH	Adjusting Principle Axis Hyperspherical
BCS	Bardeen-Cooper-Schrieffer
BEC	Bose-Einstein Condensation
CASSCF	Complete-Active-Space Self-Consistent-Field
CBS	Complete Basis Set
CMB	Cosmic Microwave Background
CRESU	Cinetique de Reaction en Ecoulement Supersonique Uniforme
DC	Delves Coordinate
DOC	Degree of Control
DOF	Degrees of Freedom
GLDP	Generalized London-Eyring Polanyi-Sato Double-Polynomial
HFS	Hyperfine Structure
LIF	Laser-Induced Fluorescence
MCP	Microchannel Plate
MOT	Magneto-Optical Trap
MQDT	Multichannel Quantum Defect Theory
MRCI	Multi-Reference Internally Contracted Configuration Interaction
NP	New Physics
PA	Photoassociation
PES	Potential Energy Surface
PI	Penning Ionization
QCT	Quasi-Classical Trajectory
QRRR	Quasi-Resonant Rotation-Rotation
QRVV	Quasi-Resonant Vibration Vibration
QT	Quantum Threshold
REMPI	Resonance-Enhanced Multi-Photon Ionization
RF	Radio Frequency

RKHS	Reproducing Kernel Hilbert-Space
RMOS	Rotating Morse Spline
SARP	Stark-Induced Adiabatic Raman Passage
SQ	Statistical Quantum
SrF	Strontium Fluoride
SW	Stark-Werner
TBR	Three-Body Recombination
WKB	Wentzel, Kramers, and Brillouin

PREFACE

Ultracold molecules and atoms offer unprecedented prospects for the controlled investigation of molecular and atomic events, which include chemical reactivity of species in the quantum realm. The propagation of systems to produce, cool, and restrain them has permitted the examination of a diverse range of chemical reactions and molecular systems at low temperatures. This text offers a detailed account of the latest progress on the theoretical and experimental aspects of cold and ultracold atoms and molecules.

Generally, cold and ultracold molecular chemistry are presumed to be disciplines dedicated to the regulation of chemical reactions involving molecules and atoms. Nevertheless, this book shows that there is a lot more than just chemical reactions in the realm of cold and ultracold chemistry. For example, the interaction between cold atoms and cold molecules result in novel properties of these species.

Furthermore, this book provides a thorough understanding of Standard Particle Physics Model (SPPM) and the examination of physics outside the standard model. The readers will notice the importance of this book with respect to systems of cold and ultracold molecules. Finally, everything contains atoms and molecules, therefore, there will always be the influence of chemistry in all material objects. A better understanding of chemistry will help us design better quantum simulators and better technologies.

There are eight chapters in the book. Chapter 1 introduces the readers with the fundamentals of cold and ultra-cold temperatures. Chapter 2 deals with the introduction of chemical and physical characteristics of cold and ultracold atoms and molecules. Chapter 3 discusses the fundamental concepts of collisions between near cold and ultracold molecules. Chapter 4 deliberates the essential concepts of collisions between cold and ultracold atoms and molecules.

Chapter 5 offers a detailed overview of the cooling mechanism induced by laser. Chapter 6 contains information about the interactions of cold and ultracold molecules in an electrostatic magnetic trap. Cold chemistry emphasizes the study of chemical reactions at temperatures 1 mK $\leq T \leq 1$ K. At these temperatures, molecules, and atom's dynamics are probably dominated through pure quantum mechanical behavior, taking captivating phenomena on chemical reactions like resonance effects. The concepts discussed briefly in Chapter 7 focusses on the physics of few single ions in ultracold bath. Finally, Chapter 8 introduces the readers to the fascinating world of quantum transport of atoms and molecules in an ultracold environment.

This book is specifically intended for graduate students involved in chemical reactions at extremely low collision energies. At lower energies, the intrinsic quantum mechanical characteristic of chemical procedures arises and becomes tangible. This book is also equally beneficial for all the readers from multidisciplinary fields.

Author

Chapter 1

Introduction to the Domain of Cold and Ultracold

CONTENTS

1.1. INTRODUCTION

This chapter gives an analysis of the present art in the field of ultracold and cold molecules. It assists as an overview of the Ultracold and Cold Molecules and defines novel prospects for technological development and fundamental research. Ultracold and cold molecules may revolutionize few-body physics and physical chemistry, provide methods for probing novel states of the quantum matter, permit accurate measurements of applied and fundamental interest, and allow quantum simulations of the condensed-matter phenomena. The ultracold molecules provide promising applications like novel platforms for quantum computing, accurate control of molecular dynamics, Bose-enhanced chemistry, and nanolithography. The discussion is centered on current theoretical and experimental work and completes with the summary of predicted open questions and future directions in this quickly growing research field (Dulieu & Gabbanini, 2009; Lemeshko et al., 2013).

Experimental work with chilled molecular gases to ultralow temperatures provides novel insights into numerous-body physics, quantum chemistry, quantum dynamics of the complex systems, and essential forces in nature. From electric and magnetic decelerators to magneto- and photoassociation (PA) to the buffer gas cooling, a range of freshly-developed experimental methods provide novel ways to physical discoveries centered on ultracold and cold molecules. The research area of cold molecules generally brings together two of three major thrusts of modern optical, molecular, and atomic physics: the ultraprecise and the ultracold. It brings together researchers from a multiplicity of fields, comprising AMO physics, quantum simulations and quantum information science, chemistry, condensed matter physics, astrophysics, and nuclear physics. To explore thrilling applications of the cold matter experimentations, it is essential to yielding large groups of molecules at temperatures below one K. The cold (1 mK to 1 K) and ultracold (less than 1 mK) temperature regimes were distinguished (Weck & Balakrishnan, 2006).

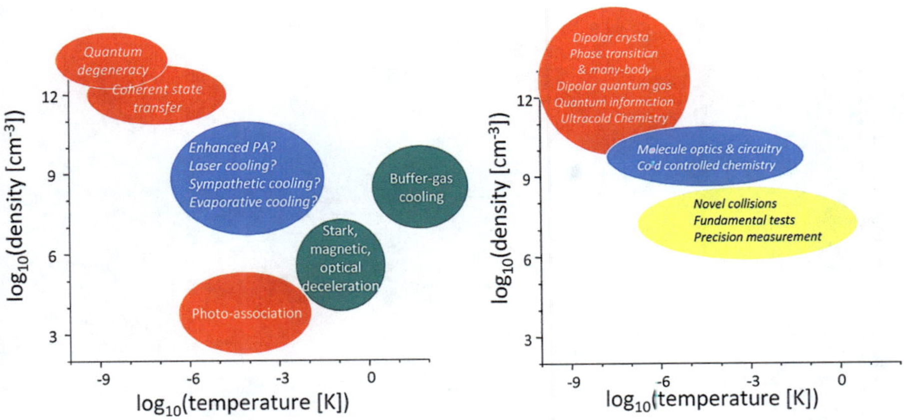

Figure 1.1. (i) The production of ultracold and cold molecules in diverse regions of temperature (T) and spatial density (n). Some technical methods that are still to be confirmed in experiments can address the significant region of n ~ 10^7 to $10^{10}/cm^3$ and T ~ 1 mK to 1μK. (ii) Applications of ultracold and cold molecules to several scientific explorations are displayed with the necessary values of T and n. The several bounds exhibited here aren't meant to be firmly applied, but instead, they aid as guidelines for the technical necessities essential for particular scientific topics.

Source: https://iopscience.iop.org/article/10.1088/1367–2630/11/5/055049

The creation of ultracold groups of atoms has transformed the field of atomic, molecular, and optical physics and produced much curiosity amongst researchers in other, usually disjoint fields. The influence of producing ultracold molecules is anticipated to be as profound and large as that normally made by the work carried out with ultracold atoms. The Molecules provide microscopic degrees of freedom (DOF) absent in the atomic gases. This provides ultracold molecular gases exclusive properties that might permit for the understanding of novel physical phenomena and might lead to findings, reaching far beyond the emphasis of customary molecular science (Salumbides et al., 2015). For instance, a Bose-Einstein condensate (BEC) of the polar molecules would signify the quantum fluid of anisotropically and strongly interacting particles and thus greatly augment the scope for understanding and applications of combined quantum phenomena. It might be used to clarify the relationship between BECs in dense liquids and dilute gases (Quemener & Julienne, 2012; Carr et al., 2009).

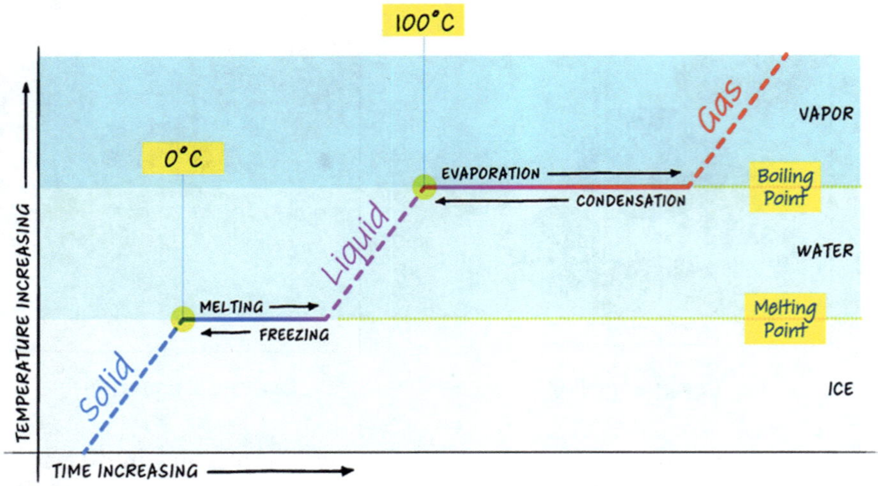

Figure 1.2. Temperature effect on the molecules (conversion from cold state to hot and vice-versa).

Source: https://sciencing.com/difference-between-hot-cold-molecules-8379015.html.

The ultracold fermion understanding is of great curiosity: electric dipole-dipole interaction might give rise to the molecular superfluid through BCS (Bardeen–Cooper–Schrieffer) pairing. The electric dipole-dipole interaction is anisotropic and long-range, and generally, it leads to novel condensed-matter phases and novel complex quantum dynamics. Groups of stuck polar molecules turned out by microwave fields might create topologically ordered states aiding excitations with the anyonic statistics (Knowlton et al., 2009; Weidemüller et al., 1998). The relationship of ultracold molecules into the chains might offer a novel system in order to model rheological phenomena encompassing the understanding of elasticity to the materials having non-classical mechanical behavior (Banerjee et al., 2012). The ultracold polar molecules stuck in the optical lattices, or adjacent mesoscopic electrical circuits, offer an encouraging platform for quantum information processing. They provide a significant benefit over neutral atoms since they have extra tunable experimental parameters: the electric dipole moment can usually be induced in the ultracold polar molecules by the static direct current electric field, and conversions amongst internal rotational states can normally be driven with the resonant microwave fields. The existence of rotationally excited states permits the likelihood to dynamically modify the dipole-dipole interactions in order to be effectively long or short-range (Doyle et al., 2004). This is besides the long times of coherence, utmost purity of a sample, and capability to hold and use molecules with the off-resonant alternating current trapping fields, e.g. incarceration in the optical lattice, all of the characteristics which they divide with neutral atoms. The selection of molecule can lead to hyperfine structure (HFS) or fine structure which can be handled together with rotational states or three-doublet states (Dulieu et al., 2011). These extra structures lead to the large inner Hilbert space, which, when joined with vibrational states in the external trap, is similar to the system of the stuck ions (Carr et al., 2009), successfully utilized for various fundamental applications. The schemes of quantum computing centered on the ultracold molecules have normally been discovered theoretically for a few years now. Therefore, there is literature developed well to build from, including everything from the molecular vapors to the hybrid mesoscopic devices (Dulieu et al., 2006) to the artificial lattice systems (Goldman et al., 2013).

1.2. WHY ULTRACOLD TEMPERATURES?

We are more aware of the Chemistry that is known and experienced every day, mostly takes place at room temperature. At the lower temperatures, molecules in the gas possess lower velocities, and therefore, for similar density, collides, and impacts are less often. Then, one might think that with the drop in temperature, the rate of reaction will follow a similar fate. This *classical* visualization is somewhat true. However, one must remember that with the drop in temperature of the gas drop, quantum mechanics generally takes over, and the phenomena unimportant at room temperature (T=298 K) will dominate the chemistry and physics of the regime. The onset of resonances is amongst these quantum phenomena (Krems, 2008; Hutson, 2007), which are usually washed out at room temperature. Other fascinating quantum phenomena are the onset behaviors of inelastic and elastic collisions, also called the Wigner threshold laws (Wigner, 1948). These laws envisage that the elastic cross-section inclines towards the constant value when the temperature reaches zero, while the inelastic cross-section upsurges with $\frac{1}{\sqrt{T}}$ (Brumer & Shapiro, 2012; Kosloff et al., 1989).

where; T →temperature of a gas.

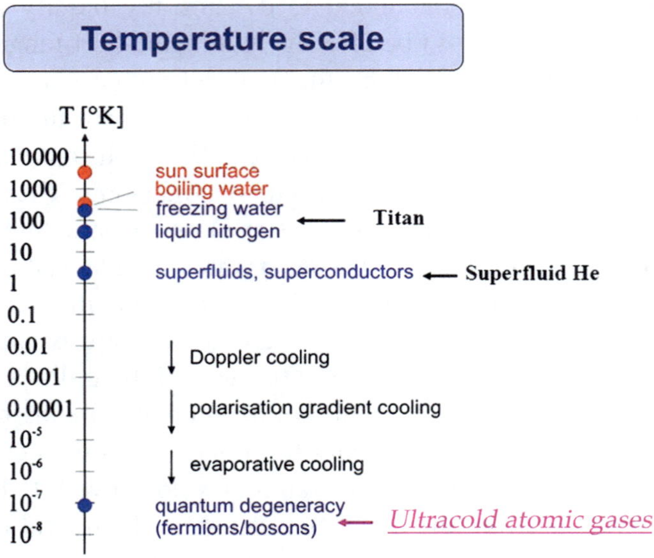

Figure 1.3. Temperature scale for diverse states of the matter.

Source: https://slideplayer.com/slide/5174385/.

Governing chemical reactions are amongst the main objectives of modern chemistry. Certainly, various methods have been proposed and developed to that end (Prokhorenko et al., 2006; Levin et al., 2015). Coherent quantum control might be the most effective, which depends on utilizing the coherence of outer laser fields in order to govern the fate of the reaction. Conversely, at room temperature, common arbitrary collisions among the molecules tempt decoherence, therefore decreasing its effectiveness.

In the ultracold system, the interaction of molecules and atoms with the external fields is usually of a similar order of magnitude as the energy of the collision, which can be used to establish accurate control over the diverse DOF of the molecule (Krems, 2008; Lepers et al., 2018; Morita et al., 2019). Therefore, ultracold temperatures trigger the novel domain for the control and handling of reactions.

This degree of control (DOC) of internal DOF can be utilized to progress quantum simulators in order to study intricate many-body issues (Chen et al., 2005; Goldman et al., 2013), to design novel quantum protocols of information (DeMille, 2002; Saffman et al., 2010), or to progress high-precision spectroscopy methods to comprehend time variation of important constants (Chin et al., 2009; Chang, 2005; DeMille et al., 2008) and physics outside the standard model (Safronova et al., 2018). The duality of wave-particle in the quantum mechanics develops that every particle also acts as the wave with the wavelength given by the following equation:

the de Broglie wavelength

where; k_B →Boltzmann constant; m →mass; h¯ →reduced Planck constant.

At the room temperature, ~10^{-11} m, m that is 10 times smaller as compared to the usual interatomic distance of the molecule ~1Å = 10^{-10} m. In order to perceive the intrinsic wave nature of the molecules at an energy scale, one would need to discover distances smaller as compared to the size of the molecule.

That is the main reason why one doesn't observe quantum effects in daily life. However, with the drop in temperature, increases and it might become of a similar order of magnitude as an interparticle distance and under suitable conditions leads to Bose-Einstein Condensation (BEC), which is unobservable otherwise. The monitoring of BEC (Quemener & Julienne, 2012; Bradley et al., 1995) transformed chemical physics, triggering a novel paradigm of ultracold physics.

Chemistry, the theme of the chapter, is appropriate at the vast temperature range. The set of principles that control chemistry isn't always the same, and they vary as the temperature (or other conditions) changes. To give the flavor of it, the sketch of suitable physical and chemical procedures characteristic of the given range of temperature is given in Figure.1.4.

The figure displays the appropriate parameter space for chemistry, concentrating on de Broglie wavelength and the temperature. For temperatures well below the ionization energy of the hydrogen atom (13.6eV), the electrons can be substituted between different atoms, triggering the creation of chemical bonds and therefore leading to the commencement of chemistry (Marinescu & You 1998; Quemener & Julienne, 2012).

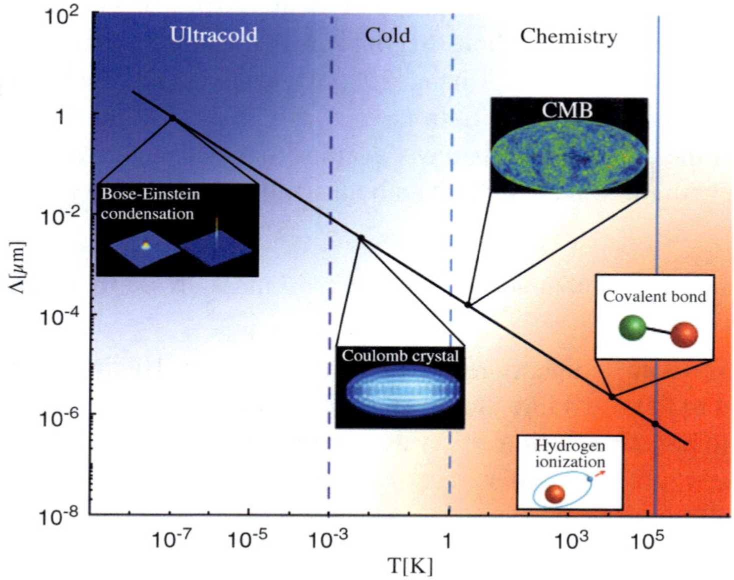

Figure 1.4. De Broglie wavelength as the function of temperature T (K) assuring parameter space for the chemistry. CMB is the abbreviation of the cosmic microwave background.

Source: https://link.springer.com/chapter/10.1007/978-3-030-55936-6_1.

At the lower temperatures, the cosmic microwave background (CMB) appears at 3 K – the lowest temperature in the entire universe. The CMB comprises microwave photons satisfying the entire universe, the remainder of the de-coupling period, the time when this universe became translucent and the photons could stream freely throughout the entire universe. Temperatures below 1 K are considered to be in a cold regime. Different phenomena take place within this region but here, the focus is on the creation of Coulomb crystals owed to its applicability to ultra cold and cold chemistry of the molecules. The Coulomb crystal is generally a structure just like crystal that appears as an outcome of the reparation amongst repulsive Coulomb interaction and the kinetic energy of ions in the existence of the trap holding these ions in the given spatial region (Hall et al., 1998; Pethick et al., 2003). The phenomenon is only feasible to perceive if there exists the friction force acting on these ions, which is normally the laser tempting the cooling of these ions. Lastly, below the mK temperature range, the ultracold region emerges. Particularly, at a temperature of 100nK, BEC can be reached. BEC is the state of matter in which the quantum state displays the macroscopic occupation, that is, the majority of the particles of a system are in a similar quantum state, triggering the appearance of the quantum effects at the macroscopic scales. It was foretold by Einstein and Bose in the 1920s, however, it took 70 long years to be witnessed in the ultracold atoms (Einstein et al., 1925; Bradley et al., 1995).

1.3. CHEMISTRY AND PHYSICS OF COLD MOLECULES

Ultracold chemistry is evolving as the new frontline of physical chemistry as discovered by the current explosion of theoretical and experimental studies on the chemical applications of ultracold and cold molecules. The terms cold and ultracold mention the translational motion of molecules, categorized by the temperature $T < 1\ K$ *or* $T < 1\ mK$, correspondingly. So far, the curiosity in ultracold and cold molecules has been stimulated largely by the variety of potential applications in diverse areas of physics, varying from quantum simulation and quantum computing of intricate many-body systems to accuracy metrology targeting at tests of important physical theories. As the novel direction, research on the cold molecules has currently started to discover crossroads with chemistry, particularly through studies of reactive and inelastic scattering in the ultracold and cold regimes. These viewpoints have driven the growth of novel experimental methods for the formation of dense groups of cold molecules, along with intense activity towards the better conceptual comprehension of chemistry at the ultralow temperatures. The conventionally close synergies between experimental and theoretical studies have now progressed the field to the point at which the cold chemistry is generally becoming a stirring reality, promising unexpected breakthroughs and discoveries (Krems, 2008; Hutson, 2007).

This special problem presents the picture of the present research in the area of cold molecules. Groups of the cold molecules, with a vision of approaching high phase-space densities near to the quantum degeneracy, are developed following approaches depending on Magneto- and photo-association of the ultracold atoms or by controlling and slowing molecular beams. Such groups are appropriate for spectroscopy of the cold molecules and novel control and trapping patterns of molecules depending on their collaboration with the external fields. This problem also presents the range of articles concentrating on cold chemistry via the scrutiny and modeling of the reactive processes or elementary collision between the cold molecules (Weck & Balakrishnan, 2006; Quemener & Julienne, 2012).

Compared with former special problems on the cold molecules, the current collection of papers outlines various novel trends. Firstly, the emphasis of field is obviously broadening to comprise ultracold and cold chemistry. The articles on this special problem demonstrate that present technology permits the creation of dense, slow molecular beams. The internal DOF and the translational motion of molecules in the beams can be handled by external magnetic and electric fields, which raises customary molecular beam experimentations to the novel level of control and precision. Secondly, studies of the cold molecular collisions generally need experimental measurements of the intermolecular interactions at low temperatures and conforming theoretical calculations for clarification. Thirdly, there is an upsurged interest in the formation of groups of the cold molecular ions, creating new possibilities for examination of the cold ion-neutral chemistry and spectroscopy of the cold ions (Lemeshko et al., 2013).

1.3.1. Ultracold and Cold Chemistry

1.3.1.1. The Standard Method

Generally, the idea of cold chemistry mentions the chemical reaction occurring at temperatures between 1 and 10^{-3}K, while the term "ultracold chemistry" is utilized for the chemical reaction happening at temperatures 10^{-3} K (Carr et al., 2009; Dulieu & Gabbanini, 2009). The definitions relate to the convention; though, they also comprise the physical meaning, as explained below.

1.3.1.2. Physical Method

In chemical physics, the appropriate physical systems are groups of molecules and atoms. The related scale of energy is given by the energy of interaction amongst colliding partners. For clarity, let's suppose atoms in this particular section. Neutral atoms interrelate through the van der Waals interaction as given below:

$$V_{\mathrm{vdW}}(r) = -\frac{C_6}{r^6}, \qquad (1)$$

where; r →interatomic distance; C_6 →van der Waals coefficient (Margenau & Kestner, 2013).

From the classical viewpoint, a collision takes place when the energy of interaction is analogous to the energy of the collision, E_k, and therefore the deviation from uniform rectilinear motion foretold by Newton's 1st law of the classical mechanics is perceived (Martinez et al., 1996; Berestetskii et al., 1982). Its quantum mechanical equivalent is given as:

$$E_K = V_{vdW} \rightarrow \frac{\bar{h}^2}{2\mu r^2} = \frac{C_6}{r^6} \tag{2}$$

Then solving the equation for r and comprising by convention the factor 1/2 produces:

$$r_{vdW} = \frac{1}{2}\left(\frac{2\mu C_6}{\bar{h}^2}\right)^{\frac{1}{4}} \tag{3}$$

where; r_{vdW} is known as van der Waals length (Jones et al., 2006). The r_{vdW} defines the typical length scale of the neutral-neutral interactions, which if replaced into the equation of quantum kinetic energy triggers:

$$E_{vdW} = \frac{\bar{h}^2}{2\mu r_{vdW}^2}\left(\frac{2\bar{h}^6}{\mu^3 C_6}\right)^{\frac{1}{2}} \tag{4}$$

Van der Waals energy. For E_k less than E_{vdW}, the scattering is subjugated by the long-range conclusion of the fundamental potential, which triggers the domain of ultracold physics. One more method of looking is that for the energies of collision E_{vdW}, only some of the partial waves goes back to scattering; thus, the scattering observables might exhibit resonance effects and the other quantum mechanical effects. E_{vdW} for numerous atom-atom systems is usually of the order of one mK, as displayed in Table 1.1. Due to this reason ultracold processes are normally those below one mK.

The ideas of van der Waals length and energy might be generalized to the long-range interaction as C_n/r_n, where $n \geq 2$. In particular, one finds;

$$r_n = \left(\frac{2\mu C_n}{\bar{h}^2}\right)^{\frac{1}{(n-2)}}; E_n = \left(\frac{\bar{h}^2}{2\mu C_n^{\frac{2}{n}}}\right)^{\frac{n}{(n-2)}} \tag{5}$$

where a factor 1/2 on the scale of length is mislaid and the relevant factor for the scale of energy too. The equations above indicate that the scales linked with the ultracold domains are strongly dependent on the nature of the system at hand. Thus, the rule E_{vdW} at 1mK isn't general.

Table 1.1. Energy Scale (mK) and Length (a0) Linked with the Ultracold frontier in Diverse Systems. The Values of C6 are Taken from (Marinescu et al., 1994), While the Coefficients of C4, Given by 1/2 of the Polarizability of the Atom (Molof et al., 1974; Ekstrom et al., 1995; Gregoire et al., 2015)

System	μ (a.m.u.)	C_6 (×10^3 a.u.)	C_4 (a.u.)	R (a_0)	E (mK)
Li-Li	3.47	1.39	-	32.4	47.55
K-K	19.98	3.81	-	64.5	2.08
Na-Na	11.49	1.47	-	44.3	7.67
Cs-Cs	66.45	6.33	-	98.9	0.27
Rb-Rb	42.73	4.43	-	81.02	0.62
Na-Na+	11.49	-	81.3	1.85×10^3	4.40×10^{-3}
Li-Li+	3.47	-	82.0	1.02×10^3	0.05
Rb-Rb+	42.73	-	160.0	4.99×10^3	1.63×10^{-4}
K-K+	19.98	-	145.0	3.25×10^3	8.20×10^{-4}
Cs-Cs+	66.45	-	199.76	6.95 10^3	5.38 10^{-5}

In the ion-atom systems or charged-neutral collaborations, the long-range interaction states as (C4/r4), where the C4 coefficient is only dependent on the polarizability of the neutral system (Margenau & Kestner, 2013). Thus, charged-neutral interactions are way more attractive as compared to neutral-neutral ones. The outcomes for the characteristic energy scale and length for ultracold physics for the ion-atom systems are given in Table 1.1. In the table, it is observed that the onset energy portraying the ultracold region for the charged-neutral systems is instead dissimilar from the one in the neutral-neutral systems. Particularly, the more striking the long-range collaboration is, the lower the threshold of energy, the larger the scale of length becomes (Chin et al., 2009; Salumbides et al., 2015).

1.3.2. Fundamental Physics and Ultracold Physics

In the current books, the reader can find the discussion of an area that is attaining more attention- the search for new physics (NP) with ultracold molecules and atoms. This novel field is dependent on utilizing the properties of ultracold molecules and atoms to carry out high accuracy spectroscopy to make novel physics models or to measure the time-dependent disparity of the essential constants of nature (Safronova et al., 2018; Ubachs et al., 2016).

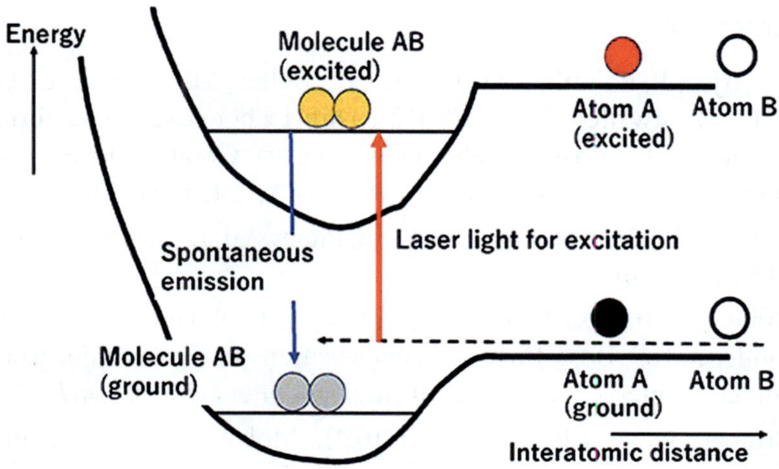

Figure 1.5. Energy states of hot and cold molecules.

Source: https://iopscience.iop.org/book/978–0-7503–3415–0/chapter/bk978–0-7503–3415–0ch4.

The external and internal DOF of molecules are readily controllable. In this particular temperature region, the collisional broadening and Doppler are negligible. Therefore, it is an exclusive playground for high-accuracy spectroscopy. Measuring the high-resolution spectra and then comparing these with the standard model likelihoods will aid to clarify the possibility of novel physics. In that method, molecules and atoms in the ultracold region are the corresponding tool for accelerator-centered experiments to increase the parameter space for the physics quest outside the standard model (Chang, 2005).

REFERENCES

1. Banerjee, J., Rahmlow, D., Carollo, R., Bellos, M., Eyler, E. E., Gould, P. L., & Stwalley, W. C., (2012). Direct photoassociative formation of ultracold KRb molecules in the lowest vibrational levels of the electronic ground state. Physical Review A, 86(5), 053428.
2. Berestetskii, V. B., Lifshitz, E. M., & Pitaevskii, L. P., (1982). Quantum Electrodynamic, 4, 1–9.
3. Bradley, C. C., Sackett, C. A., Tollett, J. J., & Hulet, R. G., (1995). Evidence of Bose-Einstein condensation in an atomic gas with attractive interactions. Physical Review Letters, 75(9), 1687.
4. Brumer, P., & Shapiro, M., (2012). Molecular response in one-photon absorption via natural thermal light vs. pulsed laser excitation. Proceedings of the National Academy of Sciences, 109(48), 19575–19578.
5. Carr, L. D., DeMille, D., Krems, R. V., & Ye, J., (2009). Cold and ultracold molecules: science, technology and applications. New Journal of Physics, 11(5), 055049.
6. Chang, Z., (2005). Chirp of the single attosecond pulse generated by a polarization gating. Physical Review A, 71(2), 023813.
7. Chen, Q., Stajic, J., Tan, S., & Levin, K., (2005). BCS–BEC crossover: From high temperature superconductors to ultracold superfluids. Physics Reports, 412(1), 1–88.
8. Chin, C., Flambaum, V. V., & Kozlov, M. G., (2009). Ultracold molecules: New probes on the variation of fundamental constants. New Journal of Physics, 11(5), 055048.
9. DeMille, D., (2002). Quantum computation with trapped polar molecules. Physical Review Letters, 88(6), 067901.
10. DeMille, D., Sainis, S., Sage, J., Bergeman, T., Kotochigova, S., & Tiesinga, E., (2008). Enhanced sensitivity to variation of m e/m p in molecular spectra. Physical Review Letters, 100(4), 043202.
11. Doyle, J., Friedrich, B., Krems, R. V., & Masnou-Seeuws, F., (2004). special issue, "ultracold polar molecules: Formation and collisions. Eur. Phys. J. D, 31(2), 149–445.
12. Dulieu, O., & Gabbanini, C., (2009). The formation and interactions of cold and ultracold molecules: New challenges for interdisciplinary physics. Reports on Progress in Physics, 72(8), 086401.

13. Dulieu, O., Krems, R., Weidemüller, M., & Willitsch, S., (2011). Physics and chemistry of cold molecules. Physical Chemistry Chemical Physics, 13(42), 18703–18704.
14. Dulieu, O., Raoult, M., & Tiemann, E., (2006). Cold Molecules: A chemistry kitchen for physicists? Journal of Physics B: Atomic, Molecular and Optical Physics, 39(19).
15. Einstein, A., (1925). Quantentheorie des einatomigen idealen gases (Zweite Abhandlung). SB Preuss. Akad. Wiss. Phys-math. Klasse, 13, 3–14.
16. Einstein, A., (2005). Quantentheorie des einatomigen idealen gases. Albert Einstein: Akademie-Vorträge: Sitzungsberichte der Preußischen Akademie der Wissenschaften 1914–1932, 237–244.
17. Ekstrom, C. R., Schmiedmayer, J., Chapman, M. S., Hammond, T. D., & Pritchard, D. E., (1995). Measurement of the electric polarizability of sodium with an atom interferometer. Physical Review A, 51(5), 3883.
18. Goldman, N., Anisimovas, E., Gerbier, F., Öhberg, P., Spielman, I. B., & Juzeliūnas, G., (2013). Measuring topology in a laser-coupled honeycomb lattice: From chern insulators to topological semi-metals. New Journal of Physics, 15(1), 013025.
19. Gregoire, M. D., Hromada, I., Holmgren, W. F., Trubko, R., & Cronin, A. D., (2015). Measurements of the ground-state polarizabilities of Cs, Rb, and K using atom interferometry. Physical Review A, 92(5), 052513.
20. Hall, D. S., Matthews, M. R., Wieman, C. E., & Cornell, E. A., (1998). Measurements of relative phase in two-component Bose-Einstein condensates. Physical Review Letters, 81(8), 1543.
21. Hutson, J. M., (2007). Feshbach resonances in ultracold atomic and molecular collisions: Threshold behavior and suppression of poles in scattering lengths. New Journal of Physics, 9(5), 152.
22. Jones, K. M., Tiesinga, E., Lett, P. D., & Julienne, P. S., (2006). Ultracold photoassociation spectroscopy: Long-range molecules and atomic scattering. Reviews of Modern Physics, 78(2), 483.
23. Knowlton, K., Rotkin-Ellman, M., King, G., Margolis, H. G., Smith, D., Solomon, G., & English, P., (2009). The 2006 California heat wave: Impacts on hospitalizations and emergency department visits. Environmental Health Perspectives, 117(1), 61–67.

24. Kosloff, R., Rice, S. A., Gaspard, P., Tersigni, S., & Tannor, D. J., (1989). Wave packet dancing: Achieving chemical selectivity by shaping light pulses. Chemical Physics, 139(1), 201–220.
25. Krems, R. V., (2008). Cold controlled chemistry. Physical Chemistry Chemical Physics, 10(28), 4079–4092.
26. Lemeshko, M., Krems, R. V., Doyle, J. M., & Kais, S., (2013). Manipulation of molecules with electromagnetic fields. Molecular Physics, 111(12, 13), 1648–1682.
27. Lepers, M., Li, H., Wyart, J. F., Quéméner, G., & Dulieu, O., (2018). Ultracold rare-earth magnetic atoms with an electric dipole moment. Physical Review Letters, 121(6), 063201.
28. Levin, L., Skomorowski, W., Rybak, L., Kosloff, R., Koch, C. P., & Amitay, Z., (2015). Coherent control of bond making. Physical review letters, 114(23), 233003.
29. Marcassa, L. G., Caires, A. R. L., Nascimento, V. A., Dulieu, O., Weiner, J., & Bagnato, V. S., (2005). Storage ring to investigate cold unidimensional atomic collisions. Physical Review A, 72(6), 060701.
30. Margenau, H., & Kestner, N. R., (2013). Theory of Intermolecular Forces: International Series of Monographs in Natural Philosophy, 18, 1–17.
31. Marinescu, M., & You, L., (1998). Controlling atom-atom interaction at ultralow temperatures by dc electric fields. Physical Review Letters, 81(21), 4596.
32. Marinescu, M., Sadeghpour, H. R., & Dalgarno, A., (1994). Dispersion coefficients for alkali-metal dimers. Physical Review A, 49(2), 982.
33. Martinez, T. J., Ben-Nun, M., & Levine, R. D., (1996). Multi-electronic-state molecular dynamics: A wave function approach with applications. The Journal of Physical Chemistry, 100(19), 7884–7895.
34. Molof, R. W., Schwartz, H. L., Miller, T. M., & Bederson, B., (1974). Measurements of electric dipole polarizabilities of the alkali-metal atoms and the metastable noble-gas atoms. Physical Review A, 10(4), 1131.
35. Morita, M., Krems, R. V., & Tscherbul, T. V., (2019). Universal probability distributions of scattering observables in ultracold molecular collisions. Physical Review Letters, 123(1), 013401.
36. Pethick, C. J., Smith, H., & Ho, T. L., (2003). Bose-Einstein condensation in dilute gases. Physics Today, 56(6), 62–63.

37. Prokhorenko, V. I., Nagy, A. M., Waschuk, S. A., Brown, L. S., Birge, R. R., & Miller, R. D., (2006). Coherent control of retinal isomerization in bacteriorhodopsin. Science, 313(5791), 1257–1261.
38. Quemener, G., & Julienne, P. S., (2012). Ultracold molecules under control! Chemical Reviews, 112(9), 4949–5011.
39. Saffman, M., Walker, T. G., & Mølmer, K., (2010). Quantum information with Rydberg atoms. Reviews of Modern Physics, 82(3), 2313.
40. Safronova, M. S., Budker, D., DeMille, D., Kimball, D. F. J., Derevianko, A., & Clark, C. W., (2018). Search for new physics with atoms and molecules. Reviews of Modern Physics, 90(2), 025008.
41. Salumbides, E. J., Schellekens, A. N., Gato-Rivera, B., & Ubachs, W., (2015). Constraints on extra dimensions from precision molecular spectroscopy. New Journal of Physics, 17(3), 033015.
42. Ubachs, W., Koelemeij, J. C., Eikema, K. S., & Salumbides, E. J., (2016). Physics beyond the standard model from hydrogen spectroscopy. Journal of Molecular Spectroscopy, 320, 1–12.
43. Weck, P. F., & Balakrishnan, N., (2006). Importance of long-range interactions in chemical reactions at cold and ultracold temperatures. International Reviews in Physical Chemistry, 25(3), 283–311.
44. Weidemüller, M., Görlitz, A., Hänsch, T. W., & Hemmerich, A., (1998). Local and global properties of light-bound atomic lattices investigated by Bragg diffraction. Physical Review A, 58(6), 4647.
45. Wigner, E. P., (1948). On the behavior of cross-sections near thresholds. Physical Review, 73(9), 1002.

Chapter 2

Chemical and Physical Characteristics of Ultra-Cold Molecules

CONTENTS

2.1. INTRODUCTION

Ultracold molecules offer exceptional prospects for the controlled examination of molecular procedures, comprising chemical reactivity in the ultimate quantum regime. The propagation of approaches to create, cool, and restrain them has permitted the examination of a miscellaneous array of molecular systems and chemical reactions at temperatures where only a single partial wave contributes. Here we propose an ephemeral account of current progress on the experimental and theoretical fronts on cold and ultracold molecules and the opportunities and challenges they deliver for an essential understanding of bimolecular chemical reaction dynamics (Chandler, 2010; Bethlem & Meijer, 2003).

Generally, the field of cold and ultracold molecules developed out of cold atom study and Bose-Einstein condensation (BEC) of atomic gases, with utmost experimental and theoretical progress arising in the atomic, molecular, and optical physics groups.

Chemists have usually been measured to hold the field, but in the latest years, a quick transferal has ensued with the viewpoint of controlled chemistry of ultracold molecules, in specific polar molecules (Krems, 2008; Jarvis et al., 2018). In this chapter, we offer a summary of contemporary progress on mutually the experimental and theoretical fronts, with specific emphasis on molecular interactions and chemical reactions in the cold and ultracold systems.

We do not try to deliver an inclusive review of the field of cold molecules but focus certain significant developments on the experimental and theoretical fronts that will assist as catalysts for future work. Numerous outstanding review articles have been published in current years (Doyle et al., 2004; Weck & Balakrishnan, 2006).

Ultracold molecules facilitate the investigation of molecular interactions and chemical reactivity at an exceptional level of detail. Reactant molecules can be produced in particular quantum states, comprising vibrational, spin, rotational, hyperfine levels, and their translational kinetic energies ranging from the nano Kelvin regime to the Kelvin regime, subject to the process selected to prepare them. Besides ultracold chemistry, cold, and ultracold molecules provide novel prospects for precision measurements, quantum information and quantum computing and quantum simulation of many-body interactions, to label a few (Jaksch & Zoller, 2005; Micheli et al., 2006; Baranov, 2008).

2.2. METHODS FOR CREATION OF COLD AND ULTRACOLD MOLECULES

Existing procedures for creating cold and ultracold molecules comprise both direct cooling of pre-existing molecules and indirect approaches like photoassociation (PA) and magneto association of ultracold atoms. Indirect procedures depend on fusing earlier cooled ultracold atoms (e.g., in a quantum degenerate gas) consuming resonant light (PA) (Bahns et al., 2000; Masnou-Seeuws & Pillet, 2001; Jones et al., 2006) or magnetic field (magneto association or Feshbach resonance method) (Donley et al., 2002; Köhler et al., 2006; Chin & Grimm, 2010). A Feshbach improved PA method in which the PA takes place in the vicinity of a Feshbach resonance has also been validated (Pellegrini et al., 2008; Krzyzewski et al., 2015). Direct methods permit the elimination of kinetic energy from pre-existing molecules employing a range of schemes. Two extensively used approaches employ the helium buffer gas cooling and Stark or magnetic (Zeeman) deceleration techniques (Schnell & Meijer, 2009; Hogan et al., 2011; Meerakker et al., 2012). These techniques have been comprehensively deliberated in current review articles by Meerakker et al. (2012) and Hutzler et al. (2012). Molecules are thermalized by elastic collisions with cold ^{3}He atoms in the buffer gas cooling technique, whereas in the Stark or Zeeman deceleration techniques, a time-varying electric or magnetic field is prudently applied to decrease the translational kinetic energy of the molecules. In the later methods, cooling arises merely in the expanding gas, whereas the electric or magnetic fields principally select a packet of molecules and amend its laboratory frame velocity without disturbing the phase space density and temperature. Patterson and Doyle (2007, 2015) and Lu et al. (2011) have protracted the buffer gas cooling process to extract a beam of cold molecules by permitting them to discharge via a hole on the side of the buffer gas cooling cell. This can be activated in the "hydrodynamically enhanced" regime or an "effusive" regime to produce bright slow beams of translationally cold molecules and radicals for cold collision experiments. The process has been validated for cold beams of O_2, CaH, Strontium Fluoride (SrF), and benzonitrile. Buffer gas-cooled molecules have also been utilized as an unceasing source of cold molecules for cavity-enhanced direct frequency comb spectroscopy and enantiomer-specific detection of chiral molecules through microwave spectroscopy (Patterson & Doyle, 2013; Patterson et al., 2013). The direct surveillance of conformational relaxation by collisions with Helium atoms in a cryogenic environment (~6

K) has also been conveyed for the 1,2-propanediol molecule (Elioff et al., 2003; Drayna et al., 2016).

Further direct methods like collisional cooling and rotating nozzle slowing and velocity filtering have been employed (Gupta & Herschbach, 1999; 2001; Strebel et al., 2010). An optical Stark deceleration arrangement utilizing optical dipole forces induced by pulsed fields has also been implemented (Fulton et al., 2004). Rempe's group has employed an optoelectrical technique for cooling and collecting polar molecules in an electric trap (Zeppenfeld et al., 2012; Zeppenfeld et al., 2016). This methodology grounded on Sisyphus cooling has been newly revealed for the CH_3F (Zeppenfeld et al., 2012) and formaldehyde (H_2CO) (Prehn et al., 2016) molecules and it embraces the great potential for precision spectroscopy and collisional studies of cold and ultracold polar molecules. In recent times, systems utilizing merged supersonic beams have been established in which a magnetic quadrupole guided beam passes at a slight angle with another supersonic beam with the purpose that the relative velocity of the collision is small (Narevicius & Raizen, 2012; Henson et al., 2012; Shagam & Narevicius, 2013). A time-of-flight mass spectrometer is then utilized to investigate the collision products. In this situation, the relative energies can be altered from the mK ranges to thermal energies by regulating the angle between the two colliding beams. The collisional or "sympathetic" cooling method in which a laser-cooled atom (usually an alkali metal or alkaline earth metal atom) collisionally thermalizes other species (atom, molecule, or ion) has also been discovered. For a method to be practicable, the rate of elastic (thermalization) collisions essentially is at least a factor of 100 larger than any inelastic loss of quenching process. This methodology is extensively employed in generating cold atomic gases along with atomic and molecular ions in ion-trap experiments) but its effectiveness for neutral molecules is yet to be verified (Willitsch et al., 2008; Rellergert et al., 2011; Chang et al., 2013).

2.3. CHARACTERISTICS OF ULTRACOLD MOLECULES

An imperative barrier that inhibits prevalent experiments on ultracold molecules is the trouble in making dense samples of molecules in particular quantum states and distinct translational kinetic energies. The lifespans of trapped molecules are constrained by collisions with inelastic and reactive scatterings that initiate trap loss. Correspondingly, the variability of molecules that can be cooled and trapped is determined by the experimental system (Stuhl & Hummon, 2014; Meerakker et al., 2012). whereas the buffer gas cooling along with magnetic trapping can technically be applied to any molecular system with an unpaired electron, its accomplishment is tangled to how rapidly the helium buffer gas can be propelled out and the ratio between elastic and inelastic collision rates. Elastic collisions result in thermalization and cooling whereas inelastic collisions initiate energy release and heating, frequently causing loss of trapped molecules. Consequently, the relative efficiency of elastic vs inelastic collision rates between the He buffer gas and the molecule is an essential parameter that envisages the efficiency of buffer gas cooling (Barry et al., 2011; Spaun et al., 2016). The identical ratio of elastic to inelastic collision rates between self-collisions of confined molecules is also a factor that defines the lifespan of trapped molecules.

If molecules are prepared in their absolute ground state, inelastic collisions are not existing (provided, no reactive scattering happens), and the lifespan is no longer constrained by collisions. Techniques based on PA and magneto association commonly produce exceedingly vibrationally excited molecules, but current STIRAP-based methods have been effective in transferring them to their absolute rovibrational ground state (Ospelkaus et al., 2010; Bergmann et al., 1998).

There is substantial attention in ultracold polar molecules. Because of strong dipole-dipole interactions, ultracold polar molecules are pliable to be regulated by an external electric field (Stuhl et al., 2012; Quéméner & Bohn, 2013). If they too have an unpaired electron as in OH, for instance, both electric and magnetic fields can be utilized to control their interaction as displayed in current calculations of OH–OH collisions by Quéméner and Bohn (2013).

These methods have appealed much to concentration recently because of the simulation of many-body interactions interceded by dipolar forces, organized studies of ultracold collisions and reactions in optical lattices, and the capability to yield novel quantum gases. Stuhl et al. (2012) have lately stated the evaporative cooling of OH down to 50 mK, and additional decline in temperature to produce a quantum degenerate gas of OH may be attainable in the near future (Idziaszek et al., 2010; Gao, 2010; 2013).

On the theoretic front, numerous substantial challenges endure both at the electronic structure theory level and for the nuclear dynamics explaining the collision process. Since collisions turn into very sensitive to slight variations in the interaction potential, precise treatment of the long-range intermolecular forces is critical for a quantitative explanation of ultracold collisions. The energy eigenvalue spectrum, particularly of the weakly bound complex that is retrieved in ultracold collisions, is also subjective by short-range forces. Consequently, a precise narrative of molecular interaction entails precise determination of both the short and long-range forces (Schnell & Meijer, 2009; Balakrishnan, 2016; Bell & Softley, 2019). Though significant advancement in electronic structure theory has been attained in current years in relating long-range forces, numerous barriers persist, specifically for molecules with an open-shell character. For high-spin open-shell systems with distinct molecular orbitals at the Hartree-Fock level, the spin-unrestricted coupled cluster with perturbative triples correction (UCCSD (T)) may be implemented. Though, open-shell systems that are not extraordinary spin and that have robust multireference character endure posing a tangible challenge. For the quantum dynamics explanation of the nuclear motion, the ultracold regime presents several supplementary problems (Ni et al., 2009; Lemeshko et al., 2013). Though one wants to contain only limited partial waves, because of the significance of long-range forces, the radial Schrödingier equation essentially is incorporated to exceptionally large separations before scattering boundary conditions can be applied. Asymptotic matching can only be completed when the interaction potential is insignificant equaled to the collision energy, which typically befalls at hundreds or, in certain circumstances, thousands of Bohr radii compared to, characteristically 15–20 bohrs in thermal energy collisions. Some contemporary improvements based on multichannel quantum defect theory (MQDT) that abridge the dynamics in the long-range are deliberated later in this chapter (Croft et al., 2011; 2012; Ruzic et al., 2013).

2.4. EXPERIMENTAL STUDIES OF COLD AND ULTRACOLD REACTIONS AND INELASTIC COLLISIONS

There is an extended history of experimental studies of chemical reaction dynamics at temperatures below 100 K, encouraged in large part because of the concern in these procedures in astrophysical environments. Crossed molecular beam methods established by Lee et al. (1969) have made the foundation of many of these experiments. In recent times, the CRESU (Cinétic de Réaction en Ecoulement Supersonique Uniforme) method (Smith, 2006; Tizniti et al., 2014) has supported the study of gas-phase chemical reactions in the 1–100 K range. Several neutral atom-molecule and molecule-molecule reactions comprising radical species have been investigated by these methods with notable results. Nevertheless, these techniques cannot reach the sub-kelvin temperatures required for cold and ultracold collisions. PA, Feshbach resonance, cryogenic helium buffer gas cooling, Stark, and magnetic deceleration, and merged beam techniques (Shagam & Narevicius, 2013; Jankunas et al., 2014; Lavert et al., 2014) endure being the favored techniques to create cold and ultracold molecules and discover their properties. To date, only laser cooling, PA, and magnetic Feshbach resonance methods were capable of producing ultracold molecules. Here we focus on certain vital results that have been conveyed in current years with specific emphasis on chemical reactivity and inelastic scattering (Jones et al., 2006; Deiglmayr et al., 2008; Ulmanis et al., 2012).

The decisive illustration of the level of control that can be attained for a chemical reaction is demonstrated in the elegant experiments on KRb molecules described by Ospelkaus et al. (2010). They utilized the isotopes ^{40}K (nuclear spin $I^{K} = 4$) and ^{87}Rb ($I^{Rb} = 3/2$) to produce KRb fermions. In this system, ultracold K and Rb atoms cooled to quantum degeneracy, are fused with a magnetic Feshbach resonance to produce ultracold KRb molecules in highly excited vibrational levels. These primarily excited KRb molecules were then transferred to the lowest energy level of the ground electronic state through a one-step two-photon transfer process like the STIRAP technique (Bergmann et al., 1998). Ospelkaus et al. (2010) provided a demonstration of the energy level diagram representing the hyperfine splitting in this scheme. The energy level separation between the two hyperfine states is an order of magnitude greater than the thermal energy of the gas, which is approximately 300 nK (Weinstein et al., 1998; Hutzler et al., 2012). They explored several different collisional routes involving KRb + KRb and K/Rb + KRb. In collisions of two similar ^{40}K^{87}Rb fermions (both in the same quantum state), *s*-wave scattering is not permitted. The lowermost permissible partial wave is a *p*-wave and at the collision energies of a few hundred nK, the collision arises below the *p*-wave centrifugal barrier. Therefore, the prime mechanism for a chemical reaction implicates tunneling through the *p*-wave barrier resulting in K_2 + Rb_2 products (the only energetically permissible product channel). The whole loss rate for this process was established to be on the order of $1.2(\pm 0.2) \times 10^{-5}$ cm^3 s^{-1} K^{-1} at 250 nK or a rate coefficient value of $3.3(\pm 0.7) \times 10^{-12}$ cm^3 s^{-1} at 250 nK for collisions of two KRb molecules prepared in either the $|-4, 1/2\rangle$ or $|-4, 3/2\rangle$ hyperfine states. Note that the rate coefficient is temperature reliant for *p*-wave collisions. When the collision partners are prepared in a 50–50 mixture of the two dissimilar hyperfine levels specified above, the limitation of *p*-wave scattering is lifted and *s*-wave scattering is permitted. Accordingly, collisions can befall deprived of any angular momentum barrier, and the resulting rate coefficient of $1.9(\pm 0.4) \times 10^{-10}$ cm^3 s^{-1} is almost two orders of magnitude greater than the indistinguishable case. The time-evolution of the KRb number density is displayed in panel (a) of Figure 2.1, whereas the temperature reliance of the rate coefficients is presented in panel (b). So, by simply flipping the nuclear spin of the collision partner, the reaction rate can be altered (controlled) by around two orders of magnitude (Lee et al., 1969; Neumark et al., 1985; Smith, 2006). Their experiment also showed that the reaction ensues through a barrier less pathway (except for the centrifugal barrier for *p*-wave scattering). In a succeeding experiment, Miranda et al. (2011) indicated that the reaction rate between two similarly created KRb molecules can be blocked by two orders of magnitude when an optical lattice is utilized to restrain the molecules in a quasi-two-dimensional, pancake-shaped geometry, with the dipoles oriented along the detention direction. In such constricted confinement geometries, only "side-by-side" collisions are permissible under the repulsive dipole-dipole interaction. Hence, Fermi statistics combined with constricted confinement in an optical lattice by an applied electric field permits controlled quantum stereo dynamics of ultracold chemical reactions (Deiglmayr et al., 2008; Ulmanis et al., 2012).

Ospelkaus et al. (2010) also studied collisions of K and Rb atoms with KRb molecules. For K + KRb collisions, the exothermic reaction resulting in the K_2 + Rb product is energetically allowable, whereas for Rb + KRb collisions the Rb_2 + K product channels are not accessible at the temperatures of the experiment. This has been in coherence with their experimental finding. Since s-wave scattering is endorsed, the reaction rate was almost $1.7(\pm 0.4) \times 10^{-10}$ cm^3 s^{-1} for K + KRb collisions, similar to the distinguishable dimer-dimer case. Nevertheless, for Rb + KRb collisions an inelastic rate coefficient of almost $0.13(\pm 0.04) \times 10^{-10}$ cm^3 s^{-1} has been conveyed. This value, about an order of magnitude lesser than that of K + KRb collisions, has been accredited to other inelastic methods comprising vibrationally excited KRb molecules present as a pollutant as the chemical reaction pathway was not accessible. The time-dependence of the KRb number density and its decay rate with either K or Rb atom density is displayed, respectively, in panels (a) and (b) of Figure 2. 2 (Schnell & Meijer, 2009; Hogan et al., 2011; Meerakker et al., 2012).

However, experiments of comparable detail as the KRb study are still to be duplicated for other schemes; several other studies have been conveyed in the last few years (Zwierlein et al., 2003; Tizniti et al., 2014).

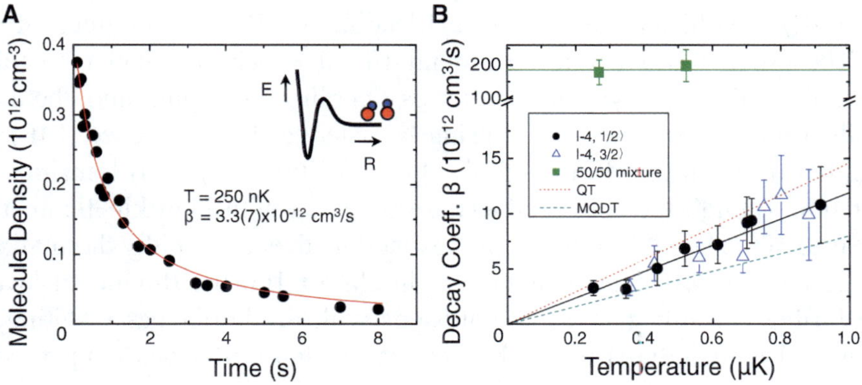

Figure 2.1. Time-dependence of the KRb molecular density due to inelastic and reactive collisions of the molecules (left panel). The KRb molecules are prepared in a single hyperfine state |−4,1/2⟩. (Right panel) Temperature dependence of the decay rate coefficient when the KRb molecules are prepared in hyperfine states |−4,1/2⟩, |−4,3/2⟩, or a 50/50 mixture of the two components. The decay rate is similar and about an order of magnitude smaller when the molecules are prepared in either of the two hyperfine states compared to the 50/50 mixture. The former occurs through p-wave collisions whereas the latter permits s-wave collisions.

Source: https://science.sciencemag.org/content/327/5967/853

Wang et al. (2013) studied ultracold collisions of ^{6}Li2 with Li, Na, and Li2 molecules. They explored that when the collision partner of $^6Li_2$ is Na or another $^6Li_2$, rapid loss arises consistently with the universal model that envisages a near-unity probability for a reaction (Idziaszek et al., 2010). In these methods, the definite nature of short-range forces is not imperative and once the short-range region is reachable inelastic loss follows at 100% efficiency. Though, when the collision partner is another 6Li atom, the collisional loss is repressed by more than an order of magnitude, specifying substantial departure from the universal performance. The lower reactivity is elucidated based on a minor density of states for the lighter system. Contrasting to the KRb experiments, the molecules were prepared in the highest vibrational state in the locality of a narrow Feshbach resonance and the inelastic loss mechanism may comprise both chemical reaction and vibrational quenching (Jochim et al., 2003; Regal et al., 2003).

Singh et al (2012) have lately reported measurements of chemical reactivity in collisions of Li and CaH leading to LiH + Ca products at ~1 K. The process is exoergic by around 10500 K. The reactant molecules were formed in a cryogenic buffer-gas cooling apparatus and the LiH product molecules were sensed using laser-induced fluorescence (LIF) by exciting them on the $A^1\Sigma(v' = 5, j' - 1) \leftarrow X^1\Sigma(v'' = 2, j'' = 0)$ transition at 401 nm. Though the LiH products are prepared with high kinetic energy because of the large exoergicity of the reaction, they are rapidly thermalized to the helium buffer gas temperature in almost 100 μs. Rotational levels are further speedily relaxed in collisions with the buffer gas resulting in rotational temperatures of ~1 K, identical to the buffer gas temperature. Though, because of low proficiency for vibrational quenching in helium collisions, the vibrational levels are relaxed on a lengthier span. This permits the surveillance of excited vibrational levels and the extent of the product vibrational distribution in cold collisions. On the other hand, because of calibration problems, product vibrational resolution has not been attained in their experiments. They conveyed an inclusive reaction rate coefficient of almost 3.6×10^{-10} $cm^3 s^{-1}$ at 1 K with a factor of 2 uncertainty (Narevicius & Raizen, 2012; Henson et al., 2012).

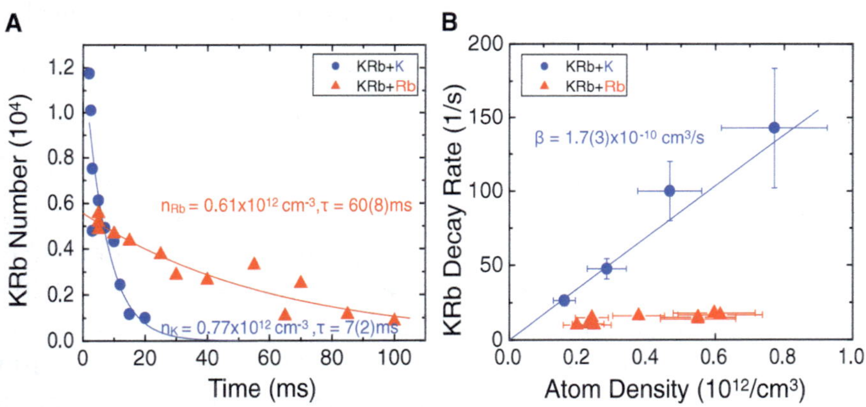

Figure 2.2. Time-dependence of the KRb molecular density due to inelastic and reactive collisions with K or Rb atoms (left panel). The rapid decay when a K atom is the collision partner is due to a chemical reaction leading to the K2+Rb product. (Right panel): KRb decay rate with atom density.

Note: The rapid decay concerning the increase in K atom density is due to an exothermic reaction leading to the K_2+Rb product. No energetically allowed reaction channels are available when Rb is the collision partner and the molecule density is unaffected by the presence of Rb atoms.

Source: https://aip.scitation.org/doi/10.1063/1.4964096]

There has also been substantial development in crossed molecular beam investigations of inelastic collisions and chemical reactions in the near-cold regime (Balakrishnan et al., 2016). In crossed molecular beam experiments with laboratory frame velocities v_1 and v_2 and reduced mass μ for the colliding species, the collision energy in the center-of-mass frame (the relative translational energy) is known in terms of the relative velocity *vr* by *E*T = 21 μvr2 =$\frac{1}{2}$ μ(v12 + v22 − 2v1v2 cos χ), where χ is the beam intersection angle. The nominal comparative collision energy is $E_T = \mu v_1^2(1 - \cos\chi)$ for $v_1 = v_2$. Thus, small beam intersection angle and truncated velocities for *both* beams are needed for achieving low collision energies in crossed beam experiments. Using this method, Lara et al. (2012) stated experimental and theoretical studies of the $S(^1D_2)$ + HD ($j = 0$) reaction at energies extending from 0.46 meV to 54.2 meV. Categorical measurements of integral cross sections for the HS + D and DS + H products and their branching ratios were matched with time-independent coupled channel quantum calculations. whereas good agreement with the experiment has been acquired for energies above 10 meV, neither the low energy resonance features (1.5–5 meV range) in the HS + D product channel nor the branching ratios were mimicked by the theoretical calculations. The incongruity was accredited in part to the limitation of a single adiabatic electronic potential energy surface (PES) for the scattering calculations. Numerous researches on inelastic collisions have been published using this method (Balakrishnan et al., 2016; Chefdeville et al., 2013). Measurements of rotational excitation cross sections in CO and O_2 caused by impacts with ortho- and para-H_2 molecules have lately been conveyed by Chefdeville et al. (2013). Ortho-H_2 (nuclear spin = 1) and para-H_2 (nuclear spin = 0) refer to two nuclear spin states of H_2 categorized by odd and even rotational levels, respectively. Several partial waves contribute at these temperatures resulting in broad resonances in the energy reliance of the cross-sections (Balakrishnan et al., 2016).

In recent times, Tizniti et al. (2014) employed the CRESU method to explore the F + H_2 reaction at 10–100 K. In the CRESU method, low temperature is attained by an isentropic expansion of a buffer gas through convergent-divergent Laval nozzles. Pure H_2 was utilized as both the buffer gas and the reactant molecule with F atoms produced by the photolysis of F_2. Rate coefficients from these measurements are compliant with quantum scattering calculations when the open-shell character of the fluorine atom and the influence of spin-orbit coupling was involved. Figure 2.3 (Tizniti et al., 2014) delivers a comparison between experimental data and various theoretical calculations. The assessments indicate that despite progress in electronic structure calculations, the different PESs lead to rate coefficients that vary by almost an order of magnitude in the 10–60 K regime. Photodetachment spectroscopy of anionic species has been extensively utilized to analyze reactive scattering resonances (Neumark, 2005). In the latest study, Kim et al. (2015) described the direct observation of reactive scattering resonances in the F + H_2 and F + D_2 reactions utilizing photodetachment of FH_2^- and FD_2^- anions. The resonance peaks have been precisely anticipated by theoretical calculations using the F + H_2 PES of Lique et al. (2011), which has also been employed in the work of Tizniti et al. (2014).

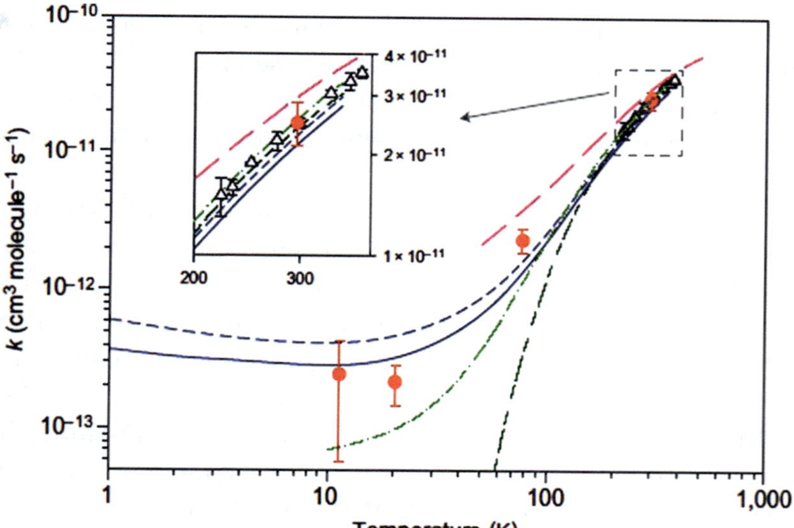

Figure 2.3. Temperature dependence of the rate coefficients for the F+H2(v =0) → HF+H reaction. The symbols represent experimental data and the various curves denote quantum dynamical calculations with different potential energy surfaces and dynamics approximations.

Source: https://www.nature.com/articles/nchem.1835.

Note: The filled red circle with error bars corresponds to the latest experimental result of Tizniti et al. (2014). The solid and dashed blue curves correspond to theoretical calculations of Tizniti et al. with and without spin-orbit coupling in the fluorine atom using the PESs of Lique et al. (2011). The long-dashed red line and green dashed-dotted lines correspond to theoretical calculations of Zhu et al. (2002) on the SW PES (Stark & Werner, 1996) and Aquilanti et al. (2012) on a modified version of the SW PES, respectively.

Narevicius and co-workers have lately made substantial improvements in using merged supersonic beams to examine the Penning ionization (PI) process over energies ranging from 10 mK to 100 K (Narevicius & Raizen, 2012; Henson et al., 2012).

The process is defined as $A^* + B \rightarrow A + B^+ + e^-$ and is an electron transfer reaction. It takes place when the ionization potential of B is smaller than the internal energy of A^*. They explored $He^* + H_2 \rightarrow He + H^+_2 + e^-$ and the influence of isotope substitution of H_2 by executing similar measurements for HD and D_2 (Henson et al., 2012; Shagam & Narevicius, 2013; Lavert et al., 2014).

It has been established that isotope substitution has a substantial influence on the ionization process with an order of magnitude difference in the reaction rate coefficient at certain collision energies.

This has been accredited to shape resonances that happen at different collision energies for the different isotopes. Figure 2.4 displays the observed isotope effect in their experimentations (Lavert et al., 2014). Jankunas et al. (2014) have lately protracted these findings to polyatomic molecules, CH_3F and NH_3/ND_3, by collisions with an electronically excited neon atom in the $^3P_2$ state. In the case of methyl fluoride, they conveyed PI cross sections in the energy range 13 μeV-4.8 meV.

However, in the ammonia system, the energy range was 8 μeV–20 meV. Appealingly, no substantial isotope effect was witnessed for the NH_3/ND_3 system. Besides PI, they also detected a reaction channel, resulting in Ne $+NH^+_2 + H + e^-$ products. The branching ratio $\frac{[NH_2^+]}{([NH_2^+]+[NH_3^+])}$ persisted around 0.3 over the complete energy regime. A parallel study of the PI of NH_3 by an electronically excited He($^3S_1$) has been stated by Jankunas et al. (2014) in the energy range 3.3 μeV10 meV. Two resonance peaks in the integral cross-section at 1.8 meV and 7.3 meV were assigned to partial waves l = 15, 16 and l = 20, 21, correspondingly (Shagam & Narevicius, 2013; Jankunas et al., 2014; Lavert et al., 2014).

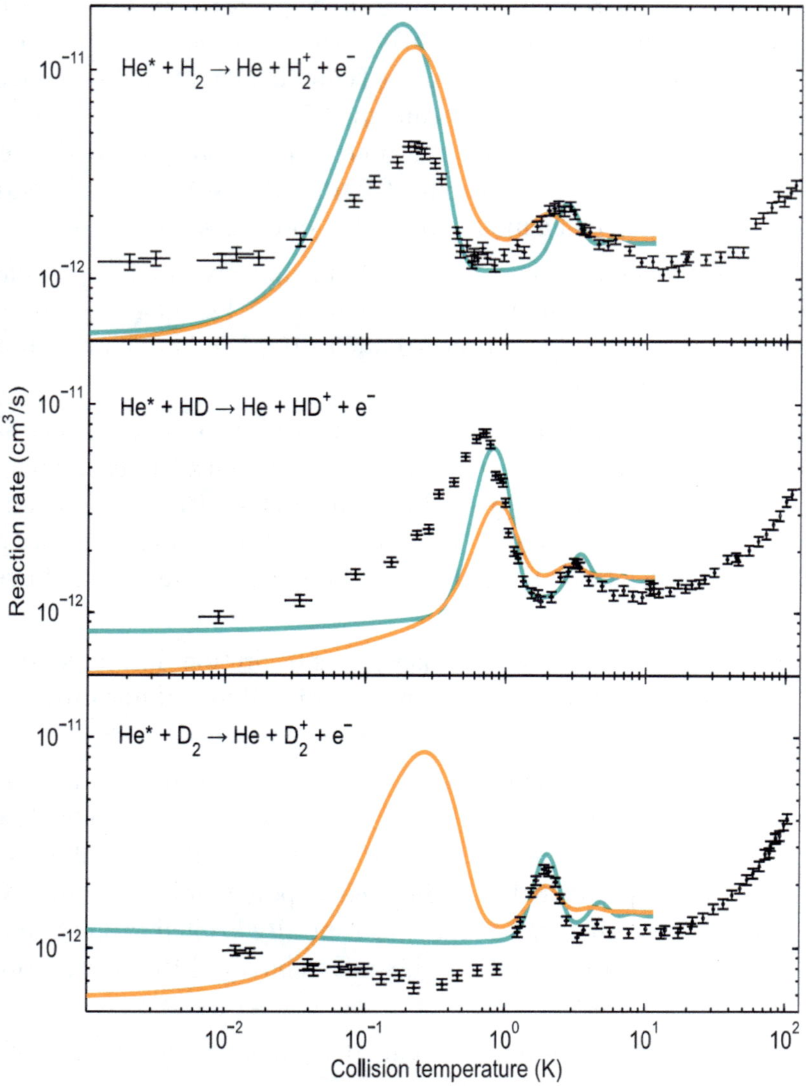

Figure 2.4. Penning ionization rates of H2 (top panel), HD (middle panel), and D2 (bottom panel) in collisions with metastable He∗(23S) as functions of the collision energy. The symbols are the experimental data and the curves correspond to different theoretical predictions with different scaling factors for the interaction potential.

Source: https://www.nature.com/articles/nchem.1857.

Various experiments of inelastic non-reactive collisions of cold and ultracold molecules with atomic and molecular counterparts have been described in the last decade. These comprise collisions of NH, CaH, CaD, and OH with He atoms established on the buffer gas cooling experiments (Weinstein et al., 1998; Hutzler et al., 2012). A mutual experimental and theoretical analysis of cold N + NH collisions has been stated by Hummon et al. (2013) at a temperature of approximately 570 mK. Atomic nitrogen and NH were co-trapped using the buffer gas cooling process and the measured trap loss rate coefficient of $9(5) \times 10^{-13}$ cm^3 s^{-1} was found to be compliant with explicit quantum scattering calculations by simial., 2007lar authors. Likewise, Stark deceleration of OH and its inelastic collisions with He, NH_3/ND_3, D_2, and NO have also been conveyed (Meerakker et al., 2005; Bochinski et al., 2003). Absolute cross sections for rotationally inelastic and spin-orbit transitions in Stark-decelerated OH with a hexapole state-selected beam of NO has been described by Kirste et al. (2012) for collision energies between 70 and 300 cm^{-1}. Results were found to be in good compliance with quantum close-coupling calculations on coupled PESs for these radical radical systems. In a succeeding study, Vogels et al. (2015) described ultrahigh-resolution measurements of differential cross sections for state-to-state inelastic transitions in Stark decelerated NO caused by Ne and Ar atoms. An angular resolution of 0.3° was accomplished for the differential cross-section. A comparative study of the NO–He system by Vogels et al. (2015) conveyed inelastic cross sections in the energy range of 13–19 cm^{-1} at an energy resolution of 0.3 cm^{-1}. Resonances were detected both in the differential and integral cross sections and outstanding agreement has been achieved with outcomes of quantum scattering calculations. Partial wave assistances to the resonances have been evaluated using partitioning of the multichannel scattering matrix, and it was reported that the computed outcomes agree with the measured data only when the resonance contribution was incorporated.

Instantaneous captivity of laser-cooled atoms and considerately cooled atomic or molecular ions offer stimulating scenarios for ultracold chemistry experiments (Hall & Aymar, 2013; Hudson, 2009). Cold reactive ion-molecule collisions have also been of considerable interest (Hall & Willitsch, 2012). In this situation, spatially ordered laser-cooled ions ("Coulomb crystals") react with cold molecules from a quadrupole-guide velocity selector. The technique has been validated for a reaction between cold Ca^+ ions and CH_3F molecules. The reaction resulting in CaF^+ ions and CH_3 radicals is exoergic by 1.05 eV (12,174 K), but it is not clear what fraction of this drives into internal rovibrational excitations of the product molecules.

The measured rate coefficient at ~1 K was explored to be about 1.3(6) × 10^{-9} cm^3 s^{-1}. This is about a factor of three greater than the room temperature value of 4.2(4) × 10^{-10} cm^3 s^{-1}. Hall and Willitsch (2012) have currently explored the reaction between laser-cooled Rb atoms and sympathetically cooled N_2^+ at a collision energy of about 20 mK. An exceedingly effective charge transferal that is determined by the internal state of the Rb atom was detected, with a four times larger rate than the Langevin rate, for Rb in the excited $(5p)^2P_{3/2}$ state. Reactions of laser-cooled Rb atoms and Ba^+ ions have also been described by Hall and Aymar (2013) lately stretched this technique to explore conformational effects in chemical reactions by studying reactions of spatially separated 3-aminophenol with cold Ca^+ ions. The spatial separation of conformers was attained through electrostatic deflection of the molecular beam. The reactivity of the *cis*-conformer was detected to be a factor of 2 higher than the trans-conformer (differentiated by the O–H bond orientation).

Researchers have stated a chemical reaction rate coefficient of about 2.0 × 10^{-10} cm^3 s^{-1} in collisions of ^{40}Ca atoms with $^{174}Yb^+$ ions restrained in a hybrid trap. While initial calculations by the authors proposed a radiative charge transfer contrivance for the large rate coefficient, succeeding calculations by researchers has shown an upper bound of 1.5 × 10^{-15} cm^3 s^{-1} for radiative transfer in the 1 mK–1 K range. So, revised measurements with an improved characterization of the products and/or calculations with better potential energy curves and non-adiabatic coupling may initiate comprehensions into the inconsistency. Rellergert et al. (2015) have also described efficient vibrational cooling of trapped $BaCl^+$ molecules by collisions with ultracold Ca atoms. For more information on charge transfer collisions and reactions of cold trapped ions (Willitsch, 2012; Balakrishna, 2016)

Three-body recombination (TBR) is one more illustration of a chemical reaction in which three atoms collide to produce a dimer and the third atom carries the excess energy. Härter et al. (2013) have currently explored explicit measurements of the vibrational, rotational, and hyperfine levels of $^{87}Rb_2$ produced in TBR in an ultracold gas of ^{87}Rb atoms. The route is Rb + Rb + Rb → Rb_2 + Rb. The experimental arrangement comprises a hybrid atom-ion trap comprising an optical dipole trap overlaid on a radiofrequency ion trap. The rubidium dimers causing by TBR were state-selectively ionized with resonance-enhanced multi-photon ionization (REMPI) method and captured in the ion trap. By measuring the Rb^+_2 ion production rate as a function of the frequency of the dipole-trap laser, Härter et al. (2013) were able to record a

REMPI spectrum of Rb2 which permitted them to draw the quantum states created in the TBR process (Wallis & Hutson, 2009; Willitsch, 2013). Results exhibited substantial rotational excitation of the Rb_2 molecule, a slightly unpredicted finding, considering that at the temperature of the experiment collisions befalls primarily in the *s*-wave regime. More advancement in experiment and theory is required in characterising product distributions in this significant class of reactions. We anticipate that the upcoming years will get numerous innovative techniques for characterising product quantum states in ultracold chemical reactions (Balakrishnan et al., 1997; Weck & Balakrishnan, 2006; Krems, 2008).

2.5. THEORETICAL DESCRIPTION OF ULTRACOLD COLLISIONS AND REACTIONS

A precise explanation of collisions and reactions in ultracold gases necessitates a fully quantum mechanical treatment. The quantum numbers ascending from spin and hyperfine levels and the effects of external electric and magnetic fields create the theoretical description of ultracold collisions, even in pure atomic gases, a significant computational challenge. For atom-molecule and molecule-molecule systems, the problem is noticeably further intricate because of inelastic and reactive channels. Time-dependent quantum-mechanical approaches however computationally more proficient than the solution of the time-independent Schrödinger equation, seem to be not well-matched for cold and ultracold collisions (Althorpe & Clary, 2003; Hu & Schatz, 2006). This is because of extended propagation times and sensitivity to absorbing potentials incorporated in time-dependent calculations. Therefore, quantum CC calculations founded on the solution of the time-independent Schrödinger equation remain to be the favored methodology for the description of cold and ultracold collisions. Consequently, numerous resourceful computational techniques based on the solution of the time-dependent Schrödinger equation conveyed in the last three decades are not employed for ultracold collisions (Jarvis et al., 2018; Schnell & Meijer, 2009).

Considerably the field of ultracold atoms and molecules would not have achieved its current status had it not been for the close compliance between experimental and theoretical results. Precise theoretical calculations of intermolecular potentials and scattering lengths for atom-atom collisions have been precarious to the initial success in cooling and trapping atoms. These comprise the understanding of atomic BEC along with molecule formation through PA and Feshbach resonance procedures (Stuhl & Hummon, 2014; Meerakker et al., 2012). Theoretical tools established in the optical, atomic, molecular, and chemical physics communities have been influential in characterising the collisional properties of cold and ultracold gases. Nevertheless, a novel additional twist is the significance of electric and magnetic field-induced energy level splitting, in addition to spatial confinement effects on collisional outcomes. Consequently, spin, and hyperfine levels, alteration in dipolar forces in the existence of an electric field, their energy splitting as a function of the magnetic field, and amendment in quantum threshold (QT) laws because of spatial confinements (abridged dimensional collisions) requires to be deliberated. The primary workings of Bohn (2000a; 200b; 2001); as well as Krems (2006); have had a substantial influence on external field control of atomic and molecular collisions. Numerous reviews have been presented in recent years, providing a detailed narrative of the theoretical approaches, and only a common background of the formalism is delivered here (Jarvis et al., 2018; Schnell & Meijer, 2009; Balakrishnan, 2016).

2.5.1. Early Studies of Inelastic Rovibrational Scattering in Atom-Diatom Systems

Theoretical investigations of rotationally and vibrationally inelastic scattering in atom-diatom systems conveyed in the late 1990s and early 2000s have been contributory in the preliminary progress in both experimental and theoretical explanation of cold molecule research (Forrey et al., 1998; Forrey et al., 1999). These findings, introduced by Balakrishnan, Forrey, and Dalgarno, and their collaborators, on the quenching of molecular rotations and vibrations in cold and ultracold gases, have revealed that the process may transpire with substantial rate coefficients at ultracold temperatures. They also established the Wigner threshold behavior of quenching cross-sections and parametrized limiting rate coefficients concerning complex scattering lengths (Balakrishnan et al., 1997). Vibrational quenching of H_2 by H and He impact was expected to rise by almost five orders of magnitude as the vibrational quantum number of the molecule was elevated from zero to 14 (Forrey et al., 1998; 1999). Further studies discovered the effect of long-range van der Waals interactions on the collisional outcome and scrutinized the role of both shape and Feshbach resonances on rotational and vibrational relaxation at low temperatures (Stuhl & Hummon, 2014; Meerakker et al.,

2012). Forrey et al. (1999) presented that quasi-resonant rotation-vibration transfers turn into noticeable, often entirely dominating the elastic collision cross-section. Quasiresonant transitions implicate instantaneous variations in rotational and vibrational quantum numbers in such a way that the modification in rotational quantum number is equivalent to an integral multiple of the variation in vibrational quantum number. For instance, $\Delta j = -2\Delta v$ or $\Delta j = -4\Delta v$ where Δj and Δv are variations in rotational and vibrational quantum numbers, dominate for He + H_2 collisions for designated j values. The collisional studies have been successively protracted to He–CO, He–O_2, Ar–H_2, He–CaH, etc. Bodo et al. (2006) and Stoecklin et al. (2002) have also studied rovibrational transitions in He–CO and other rare-gas molecule systems. Collisions comprising atom-molecular ion systems and ion-neutral molecule systems have also been explored (Weinstein et al., 1998; Hutzler et al., 2012).

2.5.2. Quasiresonant Rotational and Vibrational Transfer in Molecule-Molecule Collisions

Molecule-molecule collisions are more perplexing than atom-diatom systems when rotational and vibrational degrees of freedom (DOF) of both molecules are incorporated. Consequently, most studies of molecule-molecule systems have been constrained to the rigid-rotor case which abandons vibrational motion. It is only lately that full-dimensional quantum calculations of rovibrational transfer in H_2 + H_2 collisions were described at cold and ultracold temperatures. While H_2 is presently not experimentally pliable to ultracold collisional studies, it remains a central standard system for theoretical calculations. Furthermore, efforts are in progress, in the group of researchers, for the selective preparation of H_2 molecules in particular rotational, vibrational, and magnetic projection quantum numbers (Ni et al., 2009; Lemeshko et al., 2013). Contrasting to atom-diatom systems, molecule-molecule collisions permit simultaneous variations in rotational and vibrational quantum numbers of both molecules. As the H_2 molecule occurs in both ortho and para-modifications, ortho-para conversion does not happen in non-reactive H_2 + H_2 collisions. Therefore, calculations of para-para and ortho-ortho systems can be performed discretely. Since variations in rotational or vibrational quantum states in one molecule can be counterbalanced by like deviations in the other molecule (for identical particle collisions), molecule-molecule systems propose motivating prospects for resonant and non-resonant energy transfer at cold and ultracold temperatures. Certainly, it has been reported that in both para-para and ortho-ortho H_2 collisions, quasi-resonant rotation-rotation (QRRR) transfer takes place with high selectivity and proficiency (Quéméner et al., 2008). An illustration of such a transition is $H_2(v = 1, j = 0) + H_2(v = 0, j = 2) \rightarrow H_2(v = 1, j = 2) + H_2(v = 0, j = 0)$. In this scenario, a vibrationally excited para-H_2 molecule takes two units of rotational angular momenta from a vibrationally cold molecule. However, many other final states are possible with this preliminary pair of H_2 molecules (referred to as (1,0,0,2) where the numbers denote (v_1, j_1, v_2, j_2) levels of the two H_2 molecules), the final state is dominated by this specific transition that conserves the overall rotational angular momentum of the combined system (Balakrishnan, 2016; Bell & Softley, 2019). Figure 2.5 displays a few effective potential energy curves for this system, as functions of the intermolecular separation (Quéméner & Balakrishnan, 2009).

The H_2–H_2 interaction potential of Hinde (2008) is utilized to estimate the effective potentials. The two curves with closely degenerate asymptotes (1,0,0,2) and (1,2,0,0) correspond to the initial and final states implicated in the QRRR process.

As Figure 2. 5 demonstrates, the total internal energy is also nearly conserved in this process, the minor alteration originating from the centrifugal distortion of the $v = 0$ and $v = 1$ vibrational level because of the $j = 2$ rotational level, resulting in energy deficiency of almost 24.5 K. The process is considered to dominate inelastic cross sections for collision energies from the ultracold equal to almost 100 K.

This is shown in Figure 2. 6 in which cross-sections for the QRRR process are equated with the total inelastic quenching cross-section from a calculation that contains numerous more final states than revealed in Figure 2. 5 (Ni et al., 2009; Lemeshko wt al., 2013; Bell & Softley, 2019). It is perceived that the total inelastic cross-section is nearly completely composed of the QRRR process with the succeeding principal inelastic process that comprises pure rotational quenching, contributing less than 1%. Therefore, the high proficiency and selectivity of the transition are accredited to the conservation of internal rotational angular momentum and practically the conservation of internal energy (Weck & Balakrishnan, 2006; Krems, 2008).

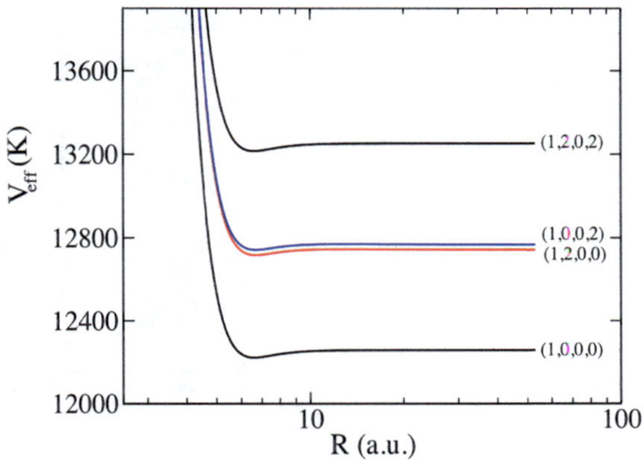

Figure 2.5. Effective potential curves for interaction between two para-H2 molecules as a function of the molecule-molecule separation. The two nearly degenerate curves correspond to the two quasi-resonant channels involved in the QRRR process discussed in the text.

Source: https://journals.aps.org/pra/abstract/10.1103/PhysRevA.90.032711.

A comparable result is detected for ortho-ortho cases, such as (1,1,0,3 → 1,3,0,1). In such case, the energy gap is marginally bigger, 42.5 K. Yet, in ortho-para H_2 collisions rotational energy exchange between the two H_2 molecules is not conceivable and the quasi-resonant transition encompasses a vibrational energy exchange, likeH_2 (v = 0, j = 1) + H_2 (v = 1, j = 0) → H_2(v = 1, j = 1) + H_2(v = 0, j = 0). This quasi-resonant vibration-vibration (QRVV) process has a minor energy gap of approximately 8.5 K but it is in orders of magnitude less effective than the QRRR process. It is driven by the isotropic part of the interaction potential, which depends only inadequately on H–H separation (Santos et al., 2011; 2013). Yang et al. (2012) have presented that both the QRRR and QRVV processes transpire for all bound vibrational levels of H_2 and that for certain H_2 vibrational levels, the process develops into more resourceful.

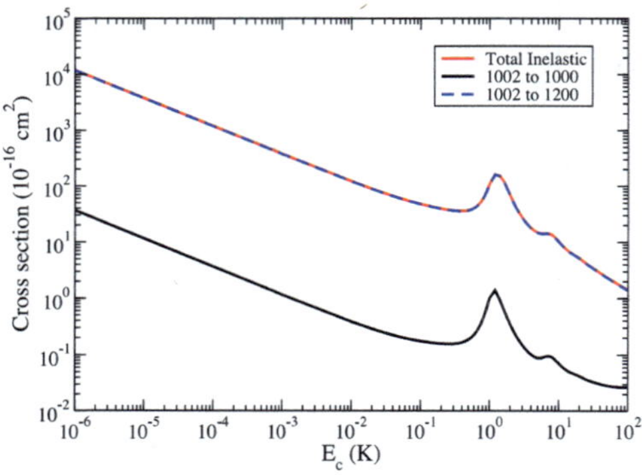

Figure 2.6. Cross sections for QRRR transition H_2(v =1, j =0)+H_2(v =0, j =2) collisions. The QRRR transition leading to the H_2(v =1, j =2)+H_2(v =0, j =0) product dominates all other inelastic channels by more than two orders of magnitude.

Source: https://aip.scitation.org/doi/abs/10.1063/1.3511699.

This is accredited to the existence of zero energy resonances near the channel thresholds of these vibrational levels. The result is comparable to that of Feshbach resonances as a function of the magnetic field but in vibration space rather than magnetic field dependence. Near-resonant processes concerning pure rotational energy transfer have been witnessed in collisions of rotationally excited HCl(j_1 = 7) with para-H_2(j_2 = 2) leading to HCl(j_1 = 9) + H_2(j_2 = 0). The energy discrepancy for the process is around 5.2 K and the process becomes prevailing for energies lower than 100 K. Widespread calculations of rotational relaxation of CO by collisions with ortho- and para-H_2 have been for CO rotational levels j = 1–40 within the rigid rotor formalism. The relaxation pathway was found to be typically subjugated by $\Delta j = -1$ transitions, for temperatures in the range of 1–1000 K. Recently, different researchers described the first full-dimensional computation of rovibrational transitions in CO by H_2 at temperatures extending from 1–100 K. Complex scattering lengths for collisions of linear and non-linear polyatomic molecules, CO_2, H_2O, and NH_3, with He has lately been stated by Yang et al. (2012). The ratio of the imaginary to real components of the scattering length, β/α, has been found to usually rise with declining rotational constant. As the energy gap, ΔE_r, for rotational transitions rises, β becomes considerably reduced than α. Remarkably, the imaginary part β seems to follow an exponential energy gap law of the form $\beta = P \exp(\Delta E_r/Q)$ for both linear and non-linear molecules, where P and Q are fit parameters.

2.5.3. Inelastic Scattering in the Presence of External Fields

Experimental investigations of ultracold atoms and molecules unvaryingly utilize external electric and/or magnetic fields. Therefore, it is required to take account of external field effects in theoretical explanations of atomic and molecular collisions in ultracold gases. This enhances substantial complications to the solution of the Schrödinger equation, and explicit calculations of ultracold molecular collisions in external fields have typically resorted to rotational and spin/hyperfine transitions in the ground vibrational level. Krems et al. (2003; 2004) performed one of the initial studies of rotational and spin-changing collisions in CaH induced by He atoms. CaH was also the first molecular system effectively cooled and trapped by the helium buffer gas cooling technique established by Doyle's group at Harvard (Doyle et al., 2004). Afterwards, numerous other molecular systems have been effectively cooled and trapped by the buffer gas cooling and Stark decelerator approaches. Experiments on OH, metastable NH, NH_3, CO, ND_3, and NO have been conveyed by many groups (Quemener & Julienne, 2012). Heavy molecules like YbF and PbO have also been exposed to buffer gas cooling and explicit quantum calculations of fine and hyperfine interactions in collisions of cold YbF and He atom in the existence of electromagnetic fields have been stated. Buffer-gas cooled YbF molecules have also endorsed sensitive measurements of the electric dipole moment of the electron. A comprehensive study of elastic and inelastic collisions of $^2\Sigma$ molecules in magnetic fields and ratios of elastic to inelastic collisions has been reported. They presented that the ratio surpasses 100 for numerous molecules comprising CN, BaF, CaF, CaH, and BaH, proposing satisfactory conditions for evaporative cooling of these molecules. The prospect of evaporative cooling of NH has been an imperative query and was theoretically addressed by Suleimanov et al. (2012). Preliminary studies supported the probability of evaporative cooling of NH through NH–NH collisions when cooled to mK temperatures by either the buffer gas technique or the Stark decelerator method. The calculations of Janssen *et al.* utilized an unconverged basis set containing only three rotational levels of NH, so the results reported are only qualitatively precise. Computations of Suleimanov et al. (2012) using the more effective total angular momentum illustration permitted a larger basis set with 7 rotational levels of the NH molecule and better convergence. These computations signify one of the largest CC calculations with approximately 19000 coupled channels. An uncoupled representation with the identical number of coupled channels would entail almost double the number of channels making such calculations computationally prohibitive. The latest calculations showed that reactive collisions through spin-exchange dominate at energies under 1 mK restraining the prospect of evaporative cooling of NH.

Frequent studies of external field effects on inelastic molecular collisions have been described. In a study conducted by Krems, it has been revealed that weakly bound van der Waals complexes can be disassociated by magnetically tuning. In this scenario, the separation arises through coupling between Zeeman levels of the bound and unbound channels and the extent of the coupling is varied by altering the external magnetic field. The prospect of sympathetic cooling of magnetically trapped polyatomic molecules by $^1S_0$-state atoms, comprising alkaline earth atoms, has been investigated. Coupled-channel quantum calculations of He + CH_2 specify that the collision-induced spin-relaxation rate is too slow to produce trap loss. Spin relaxation also arises gradually in collisions of spinpolarized CaH with Li atoms in a magnetic field regardless of the substantial anisotropy of the Li–CaH interaction, making Li atoms an auspicious cooling medium for CaH molecules. Ultracold collisions of O_2 molecules in a magnetic field have been explored which shown large rates for spin-relaxation and a dense spectrum of magnetically induced Feshbach resonances, exclusively for magnetic field strengths below 0.1 T (1000 G). An effective numerical method for locating such Feshbach resonances by computing log-derivative of the total wave function as a function of the magnetic field has been suggested by researchers (Guo, 2012; Nyman & Yu, 2013).

2.5.4. Reactive Scattering Formalism

The quantum close-coupling methodology defined above is appropriate for non-reactive collisions. To pronounce reactive collisions in an atom-diatomic molecule system, one desires to raise various sets of Jacobi coordinates suitable for the reactants and products. This coordinate transformation is a substantial concern in reactive scattering calculations. Consequently, hyperspherical coordinates are extensively employed for labeling atom-diatom exchange reactions. whereas reactive scattering for diatom-diatom systems has also been conveyed in hyperspherical coordinates, due to computational problems, to the best of our familiarity, such calculations in full dimensionality deprived of any angular momentum decoupling approximations have not been stated to date. The concept of reactive scattering in atom-diatom systems both in Jacobi and hyperspherical coordinates is comprehensively studied in various literature and is not replicated here (Althorpe & Cary, 2003; Hu & Schatz, 2006). The hyperspherical coordinates for triatomic systems implicate three internal and three external coordinates. The internal coordinates comprise a radial coordinate called the hyper-radius,, and two hyper-angles. The hyperradius is common to all reordering channels whereas the mass scaled Jacobi coordinates S_τ (atom-molecule center-of-mass distance) and s_τ (diatom distance) rely on the arrangement channel τ. The three Euler angles (α, β,γ) form the external coordinates. Different hyperspherical coordinate formalisms vary in the selection of the hyper-angles. In the Delves coordinate (DC) system, they are selected as (Θ_τ, χ_τ), where $\Theta_\tau = \arctan(s_\tau/S_\tau)$ and χ_τ is the angle between the two Jacobi vectors. In the adiabatically adjusting principle axis hyperspherical (APH) coordinates, hyper-angles are (θ, φ), where θ is given in terms of the Jacobi coordinates and φ is a constantly fluctuating kinematic angle. Note that the APH coordinates are independent of τ and they permit an impartial explanation of all three arrangement channels in an A + BC system paralleled to the DCs (Volpi & Bohn, 2002).

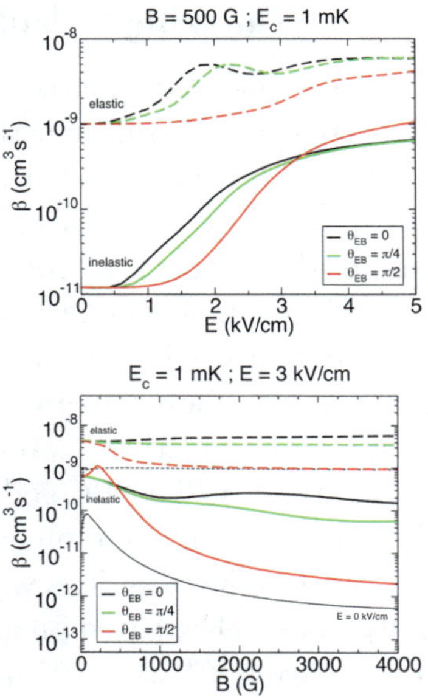

Figure 2.7. Top panel: Elastic and inelastic rate coefficients in OH+OH collisions as a function of the applied electric field for an external magnetic field strength of 500 G. The different curves for elastic and inelastic rate coefficients correspond to different relative orientation angles of the electric and magnetic fields. Bottom panel: Similar results as a function of the magnetic field for an electric field strength of 3 kV/cm. In both cases, the collision energy is 1 mK.

Source: https://journals.aps.org/pra/abstract/10.1103/PhysRevA.88.012706.

Irrespective of the choice of the hyperspherical coordinates, the elementary numerical methodology encompasses a sector-adiabatic formalism. The hyper-radius is distributed into a large number of sectors, and at the center of each sector, the total wave function is expanded in terms of hyperspherical surface functions. The expansion coefficients rely on the hyper-radius but inside a sector, they are anticipated to be independent of ρ. Coupled channel equations resulting from the Schrödingier equation with this expansion of the total wave function in terms of hyperspherical surface functions are resolved from sector to sector. Asymptotic boundary conditions are applied in Jacobi coordinates at the last sector in ρ to estimate the reactance and scattering matrices from which cross-sections and rate coefficients are computed utilizing standard expressions. The massive majority of reactive scattering calculations of atom-diatom systems at cold and ultracold temperatures utilize hyperspherical formalism.

2.5.5. Field-Free Reactive Scattering Calculations

Initially, a concise review of ultracold reactive scattering calculations stated in the absence of external fields will be carried out. Contrasting to electronic structure calculations, there are no versatile reactive scattering codes that can be applied for a wider range of molecular systems (atom-diatom, diatom-diatom, atom-triatom, etc.). whereas codes like MOLSCAT and HIBRIDON are accessible for nonreactive scattering calculations for a range of systems within different levels of estimates, such codes are not established/or extensively accessible for reactive scattering systems because of the inherent complexity in considering rearrangement collisions. A broadly used openly accessible code is the ABC code of Skouteris, Castillo, and Manolopoulos, which is compatible for tunneling dominated atom-diatom reactions. For complex-forming reactions which normally implicate deep potential energy wells, approaches based on adiabatic hyperspherical coordinates may be more applicable (Stark & Werner, 1996; Quemener & Julienne, 2012).

Theoretical studies of ultracold chemical reactions initiated with the study of the F + H_2 reaction. Their investigation, employing the Stark-Werner (SW) PES for the F + H_2 system, showed that the reaction might happen with a rate coefficient of approximately 1.25×10^{-12} cm^3/s in the ultracold limit. Succeeding calculations by researchers on newer PESs besides including spin-orbit effect in the fluorine atom have revealed equivalent reactivity though marginally less than that anticipated by the SW PES (Tizniti et al., 2014; Lique et al., 2011). Isotope effects on the F + H_2/HD/D_2 reactions and resonance enhancement of the zero-temperature rate coefficient as the hydrogen atom mass are artificially altered from H to the D atom limit has been reported. Numerous studies of chemical reactivity in other tunneling dominated reactions like Cl + H_2/HD, O + H_2, Li + HF/LiF + H, and F + HCl have been reported and deliberated in earlier reviews (Doyle et al., 2004; Weck & Balakrishnan, 2006).

Wide-ranging calculations of chemical reactivity in spinpolarized alkali metal atom-dimer collisions, with, and without vibrational excitation of the molecules, have been conveyed over the last 14 years. Atom-exchange in such systems happens without an energy barrier. Researchers explored Na + Na_2 collisions, whereas other researchers studied Li + Li_2 collisions with fluctuating degrees of vibrational excitation of the molecule. Researchers also inspected spin-polarized collisions of the K + K_2 system.

Atom-exchange reactions in collisions of heteronuclear alkali metal dimers have been extensively studied in recent years. Researchers have conducted comprehensive electronic structure calculations of energetics and reaction paths in these systems. While atom exchange in KRb + KRb $\rightarrow$ K_2 + Rb_2 is exoergic, it is not energetically permitted for several other heteronuclear alkali metal dimers in their ground rovibrational level. Certainly, researchers claimed that alkali metal dimer reactions such as 2XY $\rightarrow$ X_2 + Y_2 are energetically possible only for XY = KRb or when X = Li for singlet dimers in their ground rovibrational levels. Through comprehensive electronic structure calculations, they concluded that trimer formation reactions of the type 2XY $\rightarrow$ X_2Y + Y are also energetically not permitted for low-lying levels of the singlet state. These outcomes have been validated and also computed a minimum energy path for the KRb + KRb $\rightarrow$ K_2 + Rb_2 reaction. It has been found that the reaction implicates a barrierless path with an intermediate K_2Rb_2 tetramer at ~3000 cm^{-1} below the KRb + KRb and K_2 + Rb_2 asymptotes. Trimer formation reactions such as 2XY $\rightarrow$ X_2Y + Y are conceivable for dimers in the triplet state which are inadequately bound as well as from high-lying vibrational levels of the singlet state. For triplet dimers, 2XY $\rightarrow$ X_2 + Y_2 is exoergic except for KRb for which it is endoergic by 0.6 cm^{-1}. For KCs + KCs $\rightarrow$ K_2 + Cs_2, the exoergicity is approximately 1.0 cm^{-1} representing that the reactivity can be regulated either through an external electric field or through detention in an optical lattice. Trimer development is also energetically encouraging for alkali metal atom-dimer collisions of like species. For atom-dimer collisions like K + KRb $\rightarrow$ K_2 + Rb, or Rb + RbCs $\rightarrow$ Rb_2 + Cs, the reaction befalls through a barrierless path. Though, an obvious treatment of such systems would entail a methodology that pairs the lowest two electronic states of the trimers and the inclusion of non-adiabatic coupling between them (DeMille, 2002; Lukin, 2003; Jarvis et al., 2018).

Ultracold chemistry concerning alkali metal and alkaline earth metal atom systems has also been studied. They explored the Li + LiYb → Li_2 + Yb reaction at temperatures under 1 K using the hyperspherical coordinate methodology in APH coordinates along with statistical quantum (SQ) and MQDT approaches. It was established that the SQ and MQDT models produce equivalent results as the full quantum calculations but the product quantum state distributions from the SQ method varied meaningfully from the full quantum results.

Complex forming reactions encompassing non-alkali metal atom systems such as the O + OH → H + O_2 reaction have been explored by Quéméner et al. (2008) for OH(v = 0) and OH(v = 1) using the hyperspherical coordinate approach. Researchers have lately protracted the calculations to vibrationally excited OH (v = 2,3) and also inspected the benchmark $O(^1D)$ + H_2(v) → OH + H reaction for v = 0–2. whereas sensitivity to vibrational excitation of the OH molecule has been perceived for the former, the reactivity was revealed to be principally impervious to H_2 vibrational excitation for v = 0–2 for the latter or its isotopic $O(^1D)$ + D_2 counterpart, in the cold and ultracold regimes. This seems to specify that once the H_2O/D_2O intermediate is created, irrespective of the initial vibrational level of the molecule, its succeeding deterioration to the product or reactants is more statistical. These systems exhibit rate coefficients in the 10^{-11}–10^{-10} $cm^3\ s^{-1}$ range, around an order of magnitude greater than barrier reactions. For the latter, such as Cl + H_2 and D + H_2, vibrational excitation of H_2 is found to radically raise the reactivity, mainly due to the inclusive decrease in the effective barrier height and higher exoergicity with vibrational excitation.

A significant area in astrophysics is the ortho/para ratio in H_2 which is affected by reactive collisions of H_2 with H and H^+. The neutral reaction partakes in an energy barrier of approximately 6000 K and does not affect the ortho-para ratio at low temperatures. Nevertheless, the ionic reaction arises with no energy barrier and shows an imperative role in defining the ortho-para ratio in H_2 at low temperatures. Based on explicit close-coupling calculations, Researchers conveyed a value of 4.15×10^{-10} $cm^3\ s^{-1}$ for the conversion of ortho-H_2(j = 1) to para-H_2(j = 0) at 10 K, specifying high efficacy for the hydrogen atom exchange that drives the process. Similar calculations of D^+ + H_2 → HD + H^+ reaction using both quantum CC and the SQ methodology established and anticipated a rate coefficient value of (1.6–

2.0) × 10^{-9} cm^3 s^{-1} in the 10–100 K range. Researchers ately protracted these calculations to the ultracold regime and originated an almost temperature-independent value of ~2.0 × 10^{-9} cm^3 s^{-1} below 10^{-4} K. Certainly, the rate coefficient value endured within an order of magnitude in the temperature range 10^{-8}–100 K spanning 10 orders of magnitude, a value also precisely foretold by the Langevin model. Numerous studies of ultracold atom-diatom chemical reactions have been explored over the last 10 years. The common trend is that for barrier reactions, the rate coefficient rises with vibrational excitation till an effective barrier is no longer existent, whereas, for barrierless cases, the reactivity is an insignificant function of the vibrational quantum number, except an *s*-wave scattering resonance, is existing. Inappropriately, due to the inadequate number and variety of molecules that are amenable to cooling and trapping experiments, there is insufficient experimental data to authenticate or compare with theoretical estimates (Krems & Dalgarno, 2004; Krems, 2005).

2.5.6. External Field Effects on Reactivity

An extended history exists regarding experiments and calculations of atomic and molecular methods in external electromagnetic fields, comprising intense laser fields. Laser catalysis of chemical reactions has been stated and multi-photon ionization of atoms and molecules in intense laser fields endures to be an area of interest. The alignment of molecules in strong electric fields has been discussed and how steric effects impact reactions in external fields has been explored (Quéméner & Bohn, 2013). Nevertheless, at cold and ultracold temperatures where the incident kinetic energy is considerably reduced than perturbations induced by the external field (on the order of ~1 K), coupling of the field with the internal level structure of the molecules must be considered. Rigorous quantum theory of non-reactive collisions comprising rotational, spin, and hyperfine levels in atom-molecule and molecule-molecule collisions has been reported by Avdeenkov and Bohn (2002; 2003) and Krems and coworkers (2004). Such formalism for atom-diatom chemical reactions was principally formulated by Tscherbul and Krems (2008). They utilized an uncoupled representation for the different angular momenta and the theoretical formalism employed in the ABC code for the field-free case. By taking the LiF(v = 1, j = 0) + H → LiH + F reaction as a descriptive example, they revealed that the chemical reaction rate can be altered by orders of magnitude by introducing an external electric field. More than two orders of magnitude variation in the cross-sections were witnessed between the field-free results and those conforming to an applied electric field of 150 kV/cm for a collision energy of 1.439×10^{-3} K. Though, the results are not entirely converged on the basis set adopted. Lately, Tscherbul and Krems (2008) have reformulated the problem in the more effective total angular momentum representation and conveyed more converged outcomes (Tscherbul & Dalgarno, 2008; 2010). Cross-sections as a function of the applied field from these calculations displayed sharp electric field-induced resonances. This is demonstrated in Figure 2.8, where the peaks labeled A, B, and C stand for field-induced resonances. The inset indicates a nascent population of various HF rotational levels for multiple values of the applied electric field. Figure 2.9 exhibits the branching ratio among inelastic to reactive cross-sections as a function of the applied field. The ratio is a strong function of the applied field, signifying the prospect of controlling the collisional outcome employing an external electric field. Yet, no experimental data are existing for association with the theoretical extrapolation. Researchers have lately measured cross sections for the reverse reaction at 290 K by a crossed beam apparatus with an adjustable angle between the Li and HF beams. A magneto-optical trap (MOT) for the ultracold Li target and other sources of cold molecules like a rotating nozzle together with electrostatic guiding may perhaps permit a further decrease in temperature. The small energy behavior of the experimental outcomes specified no energy threshold for the reaction which has been verified by quantum calculations.

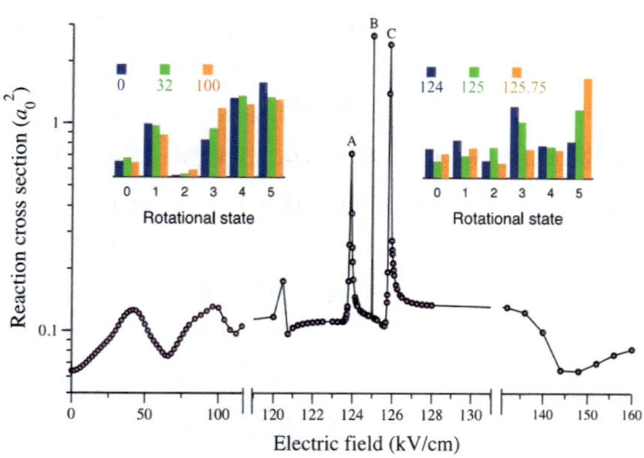

Figure 2.8. Cross-sections for LiF(v =1, j =0)+H→ Li+HF reaction at a collision energy of 0.01 cm−1 as a function of the applied electric field. The inset shows nascent rotational populations of the HF molecules for different values of the electric field.

Source: https://journals.aps.org/prl/abstract/10.1103/PhysRevLett.115.023201.

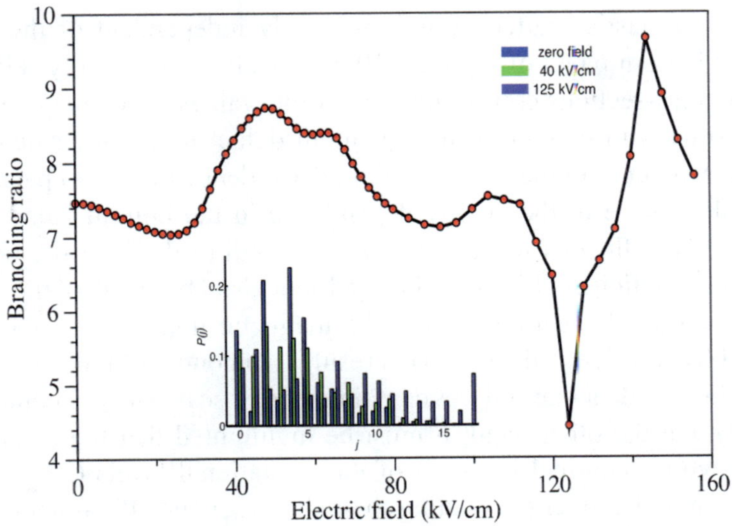

Figure 2.9. Branching ratio of non-reactive (vibrational quenching) to reactive cross sections for LiF(v =1, j =0)+H→ Li+HF reaction as a function of– the applied electric field for a collision energy of 0.01 cm^{-1}.

Source: https://journals.aps.org/prl/abstract/10.1103/PhysRevLett.115.023201.

As deliberated earlier, researchers have discussed reactive NH–NH collisions causing by a spin-exchange procedure. In such case, collisions of the two NH($^3\Sigma^-$) molecules with total electronic spin $S_{NH} = 1$ and its projection $M_{SNH} = 1$ can create dimers in high spin quintet states, $|S = 2, M_S = 2\rangle$. Spin-changing collisions can implicate either M_S of the quintet state or the total spin S resulting in singlet ($S = 0$) or triplet ($S = 1$) dimers. These spin-changing collisions produce reactive scattering as the singlet and triplet potentials show no barrier for chemical reaction. Therefore, a precise explanation of NH–NH collisions essentially comprises the influence of electronically non-adiabatic transitions. The top panel of Figure 2.10 displays the consequence of a magnetic field on elastic, inelastic, and reactive scattering in NH + NH collisions at 1 μK. The magnetic field effects become imperative for field strengths above 100 G for elastic and reactive collisions whereas the inelastic cross sections indicate strong sensitivity for magnetic field strengths as low as a tenth of a Gauss. It is comprehended that reactive scattering dominates for utmost magnetic field strengths and unveils a strong resonant enrichment for a field strength between 200 and 1000 G. When the spin-spin interaction term is ignored in the scattering calculations, the influences from reactive scattering become inconsequential compared to elastic scattering and are nearly independent of the applied field. The bottom panel of Figure 2.10 demonstrates the energy reliance of different cross-sections conforming to various values of the magnetic field. Comparisons with outcomes of a quantum defect model indicated by the dashed green curve in the magnetic field dependence in the top panel or the dashed blue curve in the energy dependence in the bottom panel specify that NH–NH collisions reveal substantial withdrawals from the universal performance anticipated by the abridged models. These calculations show that for energies beneath 100 mK, the magnetic trapping of NH will be restricted by reactive collisions. This result is in contrast to the conclusions of preceding studies that did not contain reactive scattering (Suleimanov et al., 2012). On the other hand, it must be highlighted that the results were acquired with a nominal basis set of three rotational levels ($N_{max} = 2$) and therefore are not converged relating to the basis set cut-off parameter N_{max}.

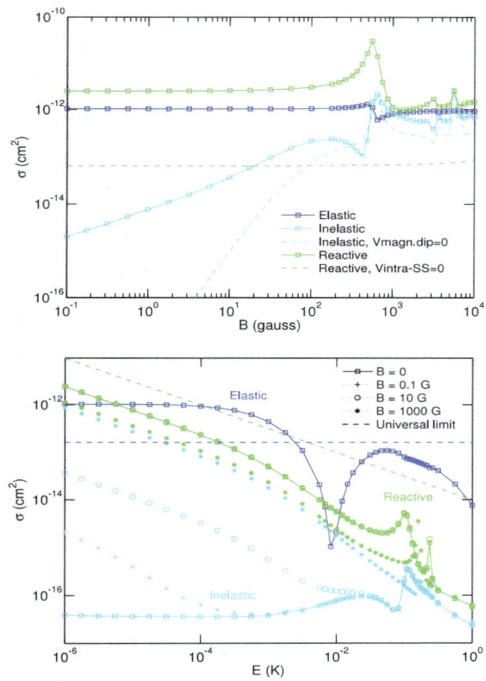

Figure 2.10. Top panel: Elastic, inelastic, and reactive cross sections as functions of the applied magnetic field for NH + NH collisions at an incident energy of 1 μK. Bottom panel: Similar results as a function of the incident collision energy for the field-free and field-induced calculations.

Source: https://journals.aps.org/prl/abstract/10.1103/PhysRevLett.110.063201.

For exoergic atom exchange reactions, for instance, $KRb + KRb \rightarrow K_2 + Rb_2$, for which the exoergicity is fairly small (~10.38(4) cm^{-1}), an electric or microwave field can be utilized to transfer the energy levels of the KRb molecules comparative to the $K_2 + Rb_2$ products (which are unaffected by an electric field except at excessive field strengths). This permits either to regulate the existing product channels or even shutdown the reactivity. Conversely, $2RbCs \rightarrow Rb_2 + Cs_2$ is endoergic by 28.7 cm^{-1} and vibrational excitation of the reactants could drive the reaction at small energies. The reactivity and accessible product channels could yet be dictated by way of shifting the rotational energy levels of the reactant molecules through an applied electric field. As lately revealed by researchers, numerous isotope exchange reactions include alkali metal and alkaline earth metal atom dimers implicate exoergicities in the cold and ultracold energy regime and feature only one open rovibrational channel these reactions can be considered devoid of any trap loss and also employed with external fields in both directions.

2.5.7. Collisions in Confined Geometries

By restraining molecules in low-dimensional geometries (two-dimensions (2D) or quasi-2D) the collisional properties and threshold performance can be varied, facilitating yet another knob to regulate and handle cold atoms and molecules. It has been revealed that the QT behavior of elastic and inelastic cross sections is determined by the dimensionality of the system. Researchers have discovered ultracold collisions in quasi-2D and 2D taking Li + Cs and $Li + Li_2$ as descriptive examples. For the later system, it has been originated that cross-sections for inelastic and reactive collisions were inhibited by detention. The threshold performance of both elastic and inelastic cross sections for the quasi-2D gas has been established to be determined by the scattering lengths of the collision partners in the confined state and the confinement length. Other researchers have investigated ultracold reactive collisions of KRb molecules in quasi-2D and 1D geometries in the MQDT framework and delivered analytical expressions for elastic and reactive rates in the universal regime. The universal regime refers to the unit possibility for reaction or inelastic quenching when the molecules are in close vicinity. It has been found that predictions for stability and evaporative cooling have been more promising for fermions as compared to bosons in the universal regime. Bose versus Fermi statistics has also a significant role in the collisional properties of KRb molecules in an optical lattice by Quéméner and Bohn (2013). They inspected ultracold collisions of KRb molecules in a 1D optical lattice and revealed that the reactive rate is improved by detention in the deficiency of an electric field. The contrary result is detected when the electric field is turned on, resulting in suppression of the reactivity. Other researchers have also explored the stability of ultracold KRb and other heteronuclear dimers of alkali metal atoms in a 1D optical lattice. In recent times, Researchers provided a formalism to evaluate ultracold collisions of reactive species in a harmonic trap interacting through an isotropic potential with instructive outcomes for KRb and LiCs molecules (Idziaszek et al., 2010).

Numerous experimental investigations of ultracold molecule collisions and reactions in restrained geometries have been conveyed recently. Miranda et al. (2011) presented that the reaction rate between two similarly created fermionic KRb molecules in an optical lattice with a quasi-2D pancake-shaped geometry is intimidated by two orders of magnitude. Lattice confined polar molecules are model systems to explore many-body quantum effects and to comprehend lattice spin models for examining quantum magnetism. Researchers validated *in situ* formations of ultracold fermionic KRb molecules by instantaneously loading a bosonic Rb atom and a fermionic K atom at every lattice site of a 3D optical lattice. Ground state KRb molecules, one per lattice site, were created by magneto association of the two atoms trailed by optical transfer, attaining a 25% filling rate. This "quantum synthesis" of low entropy gas of polar molecules in a 3D optical lattice delivers a perfect system for investigations of quantum many-body effects interceded over dipole-dipole interaction (Jankunas et al., 2014; Lavert et al., 2014).

2.6. QUANTUM DEFECT THEORIES AND CAPTURE/ LANGEVIN METHODS

Full close-coupling calculations of ultracold collisions comprising external field effects are a challenging task. Consequently, there is extreme attention in evolving abridged models founded on long-range theories. whereas Langevin-type models can deliver qualitative approximations of reaction rates for systems with strong long-range interaction, their legitimacy is uncertain at ultracold energies. In recent times, there has been significant concern in creating formalisms based on MQDT. Latest works of Idziaszek and Julienne (2010), and Gao (2010) have explored the development of universal models for ultracold reactions within the MQDT framework. The universal model of Idziaszek and Julienne (2010) the quantum reflection model of Kotochigova, and the QT models of Quéméner et al. (2013) has been effective in appraising the overall reaction rate coefficients for KRb + KRb and K + KRb reactions conveyed by the JILA group.

The elementary idea of MQDT is to resolve the Schrödinger equation precisely for the long-range potential and use an estimated explanation at the short-range. In the case of the QT model, the reaction rate coefficients are systematically assessed from the threshold laws that are pertinent in the existence of an electric field joined with a classical capture model. Although these methodologies envisage the total reaction rate coefficient, they do not produce product resolved reaction rates or differentiate between many product channels if they exist. Since the resolution of product quantum states is a vital feature of state-to-state and controlled chemistry, extension of these methodologies to harvest product quantum state resolution will be a key step (Gao, 2010).

The effectiveness of the MQDT method for cold collisions curtails the large discrepancy in energy scales for the relative motion in the incident channels and the deep potential wells that manipulate rovibrational transitions and chemical transformations. Hence, energy disparities over numerous orders of magnitude in the initial channels (say 1 μK-1 K) are insignificant compared to the depth of the potential wells ranging from 100 to 50000 K for numerous reactions.

These two regimes can be defined contrarily and the precision of the method enhanced by treating the short-range dynamics more accurately. Lately, Croft et al. (2011; 2013) have protracted the MQDT formalism to include the short-range interaction.

Their methodology implicates full CC calculations of the short-range dynamics at one or a few designated values of energy and magnetic field strength and the MQDT formalism at long-range. A short-range *K*-matrix is acquired by matching the solution of CC equations to MQDT reference functions calculated from the long-range potential at some intermediate distance, R_{match}, which is considerably smaller than the asymptotic matching distance R_{max}. The intermediate matching distance is selected such that the interchannel couplings are insignificant for distances beyond R_{match}.

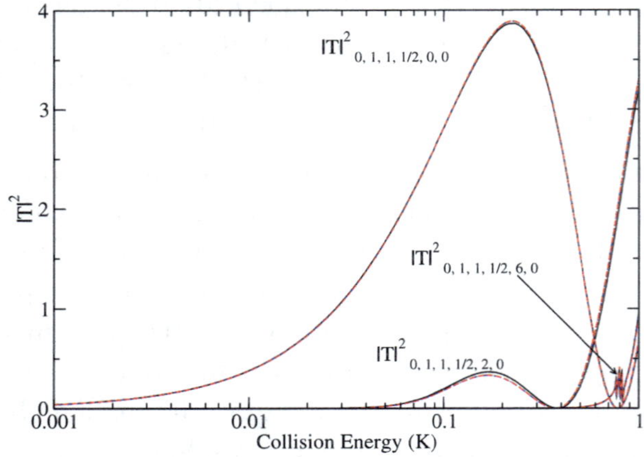

Figure 2.11. Squares of diagonal T-matrix elements Tn, j, m_j, ms_{Li}, L, M_L for Li + NH collisions in the incoming channels for m_j =+1 and L =0, 2 and 6 at B =10 G.

Source: https://journals.aps.org/pra/abstract/10.1103/PhysRevA.87.032710.

Values obtained from full CC calculations (solid, black) and MQDT using optimized reference functions for R_{match} =6.5 Å both with (dotted-dash, blue) and without (dashed, red) interpolation of the short-range K-matrix. L =4 is not shown because it obscures the resonant feature for L =6.

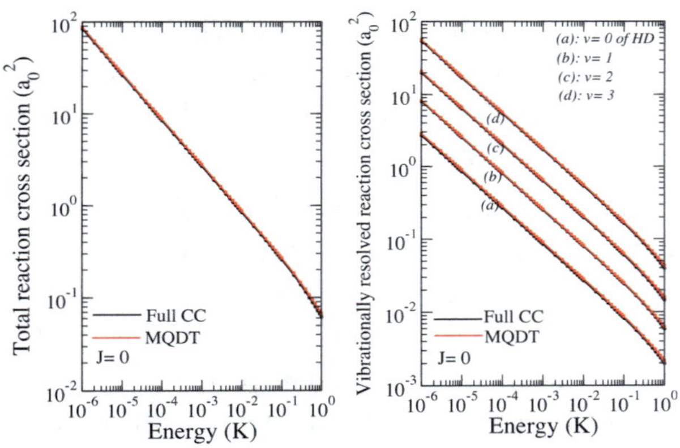

Figure 2.12. Cross-sections for D+H2(v =3, j =0) → HD(v)+H reaction as a function of the incident kinetic energy. The left panel depicts reaction cross-section summed over all v levels of HD; the right panel presents vibrational level resolved reaction cross sections for the HD product.

Source: https://pubmed.ncbi.nlm.nih.gov/27782475/.

The MQDT formalism can then be utilized to change the short-range *K*-matrix to the physical S-matrix from which cross-sections, as well as rate coefficients, can be calculated. By prudently indicating the MQDT reference functions, the short-range *K*-matrix can be made inadequately reliant on energy and external field strengths. An interpolation of the short-range *K*-matrix in energy and magnetic field strength could later be employed to radically decrease the computational cost of the CC calculations. Croft et al. (2011, 2013) have lately validated this methodology for inelastic spin-changing collisions in Mg + NH and Li + NH systems in the existence of an external magnetic field but entreating a rigid rotor approximation for the molecules. They presented that with suitable selections for the short-range matching distance and a reference potential, precise results for a spin and hyperfine transitions can be achieved. Figure 2.11 displays the energy dependence of the squares of diagonal *T*-matrix elements for Li + NH collisions as functions of the collision energy from full CC calculations and the MQDT formalism. This is in conformance with the agreement, even near the resonance features at 0.7 K.

Currently, researchers have used a similar methodology for rovibrational transitions in $H_2 + H_2$ collisions in full dimensionality. In their method, the MQDT formalism established by Ruzic et al. (2013). has been adopted and applied to QRRR and vibration-vibration transfer in para-para, ortho-ortho. It was revealed that the CC-MQDT method produces results in close conformance with the full CC calculations. Moreover, a short-range *K*-matrix estimated at a collision energy of 1 μK was found to produce precise results in the complete Wigner threshold regime encompassing 10 mK. Therefore, it seems that for the utmost part, one may limit the time-consuming CC calculations to just the short-range at one or a small number of values of collision energies and then handle the energy dependency and long-range dynamics via the MQDT formalism.

Researchers have currently protracted the technique to reactive scattering by explicitly considering the product channels. Since reactive scattering calculations are more appropriately formulated in hyperspherical coordinates, and the MQDT formalism in Jacobi coordinates, unusual precaution has to be taken in matching solutions of the CC equations to MQDT reference functions. They employed this formalism in the ABC reactive scattering code taking $D + H_2(v, j) \rightarrow HD(v', j') + H$ reaction as a clarifying example. Figure 2.12 displays the energy dependency of the total reaction cross-section (left panel) and vibrational level resolved cross sections for the HD product (right panel) as functions of the incident collision energy estimated utilizing the full CC method and the CC-MQDT method. In such a case, the short-range matching distance has been preferred to be 15 a_0 for the CC-MQDT calculation equaled to an asymptotic matching distance of 100 a_0 for the full CC calculation. The results specify that this is a favorable methodology for ultracold chemical reactions with full resolution of product rovibrational quantum states.

Since the CC calculations can be introverted to a comparatively short value of the hyper-radius outside which coupling to reactive channels can be abandoned, the computational savings are substantial compared to full CC calculations. Moreover, as in the nonreactive case, the energy (and field) independence of the short-range *K*-matrix can be oppressed to accelerate the calculations. Whether the method will be appropriate to systems with strong anisotropic interactions is yet to be validated through the current work of some researchers on inelastic scattering seems to be auspicious. The SQ method of a researcher has also been perceived consideration in recent years to designate low-temperature capture reactions. It has been applied to numerous complex-forming reactions and hydrogen atom exchange in $H^+ + H_2$ and $D^+ + H_2$ reactions with sensible achievement. As deliberated earlier, it has also been effectively applied to the Li + LiYb reaction. Researchers have currently instigated a statistical capture model for ultracold reactions in external fields and applied it to KRb + KRb $\rightarrow$ $K_2 + Rb_2$ and K + $Rb_2 \rightarrow$ Rb + KRb reactions. The performance of the capture model in the multiple-partial-wave regime ($\sim$1 K) has been evaluated recently for the Li + CaH$\rightarrow$ LiH + Ca chemical reaction and good conformance was achieved between the measured (Singh et al., 2012) and calculated values. An adiabatic capture theory for cold atom-molecule collisions with strong anisotropic interaction has also been developed by researchers. The method was magnificently applied to PI reaction in He (^3S)+HD collisions conveyed by Lavert-Ofir et al. (2014). We anticipate these approaches to gain further popularity in the impending years.

REFERENCES

1. Althorpe, S. C., & Clary, D. C., (2003). Quantum scattering calculations on chemical reactions. Annual Review of Physical Chemistry, 54(1), 493–529.
2. Aquilanti, V., Mundim, K. C., Cavalli, S., De Fazio, D., Aguilar, A., & Lucas, J. M., (2012). Exact activation energies and phenomenological description of quantum tunneling for model potential energy surfaces. The F+ H_2 reaction at low temperature. Chemical Physics, 398, 186–191.
3. Arthurs, A. M., & Dalgarno, A., (1960). The theory of scattering by a rigid rotator. Proceedings of the Royal Society of London. Series A. Mathematical and Physical Sciences, 256(1287), 540–551.
4. Avdeenkov, A. V., & Bohn, J. L., (2002). Collisional dynamics of ultracold OH molecules in an electrostatic field. Physical Review A, 66(5), 052718.
5. Avdeenkov, A. V., & Bohn, J. L., (2003). Linking ultracold polar molecules. Physical Review Letters, 90(4), 043006.
6. Bahns, J. T., Gould, P. L., & Stwalley, W. C., (2000). Formation of cold (T 1 K) molecules. Advances In Atomic, Molecular, and Optical Physics, 42, 171–224.
7. Balakrishnan, et al., (2016). S., Stoecklin, T., Bergeat, A., Hickson, K. M., Naulin, C., & Costes, M., (2012). Appearance of low energy resonances in CO–para-H_2 inelastic collisions. Physical Review Letters, 109(2), 023201.
8. Balakrishnan, N., (2016). Perspective: Ultracold molecules and the dawn of cold controlled chemistry. The Journal of Chemical Physics, 145(15), 150901.
9. Balakrishnan, N., Dalgarno, A., & Forrey, R. C., (2000). Vibrational relaxation of CO by collisions with 4 He at ultracold temperatures. The Journal of Chemical Physics, 113(2), 621–627.
10. Balakrishnan, N., Kharchenko, V., Forrey, R. C., & Dalgarno, A., (1997). Complex scattering lengths in multi-channel atom-molecule collisions. Chemical Physics Letters, 280(1, 2), 5–9.
11. Balakrishnan, N., Quéméner, G., Forrey, R. C., Hinde, R. J., & Stancil, P. C., (2011). Full-dimensional quantum dynamics calculations of H_2–H_2 collisions. The Journal of Chemical Physics, 134(1), 014301.
12. Baranov, M. A., (2008). Theoretical progress in many-body physics with ultracold dipolar gases. Physics Reports, 464(3), 71–111.

13. Barry, J. F., Shuman, E. S., & DeMille, D., (2011). A bright, slow cryogenic molecular beam source for free radicals. Physical Chemistry Chemical Physics, 13(42), 18936–18947.
14. Barry, J. F., Shuman, E. S., Norrgard, E. B., & DeMille, D., (2012). Laser radiation pressure slowing of a molecular beam. Physical Review Letters, 108(10), 103002.
15. Bell, M. T., & Softley, T. P., (2009). Ultracold molecules and ultracold chemistry. Molecular Physics, 107(2), 99–132.
16. Bergmann, K., Theuer, H., & Shore, B. W., (1998). Coherent population transfer among quantum states of atoms and molecules. Reviews of Modern Physics, 70(3), 1003.
17. Bethlem, H. L., & Meijer, G., (2003). Production and application of translationally cold molecules. International Reviews in Physical Chemistry, 22(1), 73–128.
18. Bochinski, J. R., Hudson, E. R., Lewandowski, H. J., Meijer, G., & Ye, J., (2003). Phase space manipulation of cold free radical OH molecules. Physical Review Letters, 91(24), 243001.
19. Bodo, E., & Gianturco, F. A., (2006). Collisional quenching of molecular ro-vibrational energy by He buffer loading at ultralow energies. International Reviews in Physical Chemistry, 25(3), 313–351.
20. Bohn, J. L., (2000a). Cold collisions of O_2 with helium. Physical Review A, 62(3), 032701.
21. Bohn, J. L., (2000b). Molecular spin relaxation in cold atom-molecule scattering. Physical Review A, 61(4), 040702.
22. Bohn, J. L., (2001). Inelastic collisions of ultracold polar molecules. Physical Review A, 63(5), 052714.
23. Chandler, D. W., (2010). Cold and ultracold molecules: Spotlight on orbiting resonances. The Journal of Chemical Physics, 132(11), 110901.
24. Chang, Y. P., Długołęcki, K., Küpper, J., Rösch, D., Wild, D., & Willitsch, S., (2013). Specific chemical reactivities of spatially separated 3-aminophenol conformers with cold Ca^+ ions. Science, 342(6154), 98–101.
25. Chefdeville, S., Kalugina, Y., Van De, M. S. Y., Naulin, C., Lique, F., & Costes, M., (2013). Observation of partial wave resonances in low-energy O_2–H_2 inelastic collisions. Science, 341(6150), 1094–1096.
26. Chefdeville, S., Stoecklin, T., Naulin, C., Jankowski, P., Szalewicz, K.,

Faure, A., & Bergeat, A., (2015). Experimental and theoretical analysis of low-energy CO + H_2 inelastic collisions. The Astrophysical Journal Letters, 799(1), L9.

27. Chin, C., & Grimm, R., (2010). Feshbach resonances in ultracold gases. In: Julienneen, P., & Tiesinga, E., (eds.), Rev. Mod. Phys. (Vol. 82, No. 1225, p. 71).
28. Croft, J. F., & Hutson, J. M., (2013). Multichannel quantum defect theory for cold molecular collisions with a strongly anisotropic potential energy surface. Physical Review A, 87(3), 032710.
29. Croft, J. F., Hutson, J. M., & Julienne, P. S., (2012). Optimized multichannel quantum defect theory for cold molecular collisions. Physical Review A, 86(2), 022711.
30. Croft, J. F., Wallis, A. O., Hutson, J. M., & Julienne, P. S., (2011). Multichannel quantum defect theory for cold molecular collisions. Physical Review A, 84(4), 042703.
31. De Miranda, M. H. G., Chotia, A., Neyenhuis, B., Wang, D., Quéméner, G., Ospelkaus, S., & Jin, D. S., (2011). Controlling the quantum stereo dynamics of ultracold bimolecular reactions. Nature Physics, 7(6), 502–507.
32. Deiglmayr, J., Grochola, A., Repp, M., Mörtlbauer, K., Glück, C., Lange, J., & Weidemüller, M., (2008). Formation of ultracold polar molecules in the rovibrational ground state. Physical Review Letters, 101(13), 133004.
33. DeMille, D., (2002). Quantum computation with trapped polar molecules. Physical Review Letters, 88(6), 067901.
34. Dong, W., Mukherjee, N., & Zare, R. N., (2013). Optical preparation of H_2 rovibrational levels with almost complete population transfer. The Journal of Chemical Physics, 139(7), 074204.
35. Donley, E. A., Claussen, N. R., Thompson, S. T., & Wieman, C. E., (2002). Atom-molecule coherence in a Bose-Einstein condensate. Nature, 417(6888), 529–533.
36. Dos, S. S. F., Balakrishnan, N., & Lepp, S., (2011). In: Quéméner, Forrey, R. C., Hinde, R. J., & Stancil, P. C., (eds.), J. Chem. Phys. (Vol. 134, p. 214303).
37. Dos, S. S. F., Balakrishnan, N., Forrey, R. C., & Stancil, P. C., (2013). Vibration-vibration and vibration-translation energy transfer in H2-H2 collisions: A critical test of experiment with full-dimensional quantum dynamics. The Journal of Chemical Physics, 138(10), 104302.

38. Doyle, J., Friedrich, B., Krems, R. V., & Masnou-Seeuws, F., (2004). Quo Vadis, cold molecules? The European Physical Journal D-Atomic, Molecular and Optical Physics, 31(2), 149–164.

39. Drayna, G. K., Hallas, C., Wang, K., Domingos, S. R., Eibenberger, S., Doyle, J. M., & Patterson, D., (2016). Direct time-domain observation of conformational relaxation in gas-phase cold collisions. Angewandte Chemie International Edition, 55(16), 4957–4961.

40. Dulieu, O., & Gabbanini, C., (2009). The formation and interactions of cold and ultracold molecules: New challenges for interdisciplinary physics. Reports on Progress in Physics, 72(8), 086401.

41. Dürr, S., Volz, T., Marte, A., & Rempe, G., (2004). Observation of molecules produced from a Bose-Einstein condensate. Physical Review Letters, 92(2), 020406.

42. Elioff, M. S., Valentini, J. J., & Chandler, D. W., (2003). Subkelvin cooling no molecules via "billiard-like" collisions with argon. Science, 302(5652), 1940–1943.

43. Forrey, R. C., Balakrishnan, N., Dalgarno, A., Haggerty, M. R., & Heller, E. J., (1999). Quasiresonant energy transfer in ultracold atom-diatom collisions. Physical Review Letters, 82(13), 2256, 2657.

44. Forrey, R. C., Balakrishnan, N., Kharchenko, V., & Dalgarno, A., (1998). Feshbach resonances in ultracold atom-diatom scattering. Physical Review A, 58(4), R2645.

45. Forrey, R. C., Kharchenko, V., Balakrishnan, N., & Dalgarno, A., (1999). Vibrational relaxation of trapped molecules. Physical Review A, 59(3), 2146.

46. Fulton, R., Bishop, A. I., & Barker, P. F., (2004). Optical Stark decelerator for molecules. Physical Review Letters, 93(24), 243004.

47. Gao, B., (2010). Universal model for exoergic bimolecular reactions and inelastic processes. Physical Review Letters, 105(26), 263203.

48. Gao, B., (2013). Quantum-defect theory for - 1/r 4-type interactions. Physical Review A, 88(2), 022701.

49. Greiner, M., Regal, C. A., & Jin, D. S., (2003). Emergence of a molecular Bose-Einstein condensate from a fermi gas. Nature, 426(6966), 537–540.

50. Guo, H., (2012). Quantum dynamics of complex-forming bimolecular reactions. International Reviews in Physical Chemistry, 31(1), 1–68.

51. Gupta, M., & Herschbach, D., (1999). A mechanical means to produce intense beams of slow molecules. The Journal of Physical Chemistry A, 103(50), 10670–10673.
52. Gupta, M., & Herschbach, D., (2001). Slowing and speeding molecular beams by means of a rapidly rotating source. The Journal of Physical Chemistry A, 105(9), 1626–1637.
53. Hall, F. H., & Willitsch, S., (2012). Milli-Kelvin reactive collisions between sympathetically cooled molecular ions and laser-cooled atoms in an ion-atom hybrid trap. Physical Review Letters, 109(23), 233202.
54. Hall, F. H., Aymar, M., Raoult, M., Dulieu, O., & Willitsch, S., (2013). Light-assisted cold chemical reactions of barium ions with rubidium atoms. Molecular Physics, 111(12, 13), 1683–1690.
55. Härter, A., Krükow, A., Deiß, M., Drews, B., Tiemann, E., & Denschlag, J. H., (2013). Population distribution of product states following three-body recombination in an ultracold atomic gas. Nature Physics, 9(8), 512–517.
56. Henson, A. B., Gersten, S., Shagam, Y., Narevicius, J., & Narevicius, E., (2012). Observation of resonances in penning ionization reactions at sub-kelvin temperatures in merged beams. Science, 338(6104), 234–238.
57. Herschbach, D., (2009). Molecular collisions, from warm to ultracold. Faraday Discussions, 142, 9–23.
58. Hinde, R. J., (2008). A six-dimensional H_2–H_2 potential energy surface for bound state spectroscopy. The Journal of Chemical Physics, 128(15), 154308.
59. Hogan, S. D., Motsch, M., & Merkt, F., (2011). Deceleration of supersonic beams using inhomogeneous electric and magnetic fields. Physical Chemistry Chemical Physics, 13(42), 18705–18723.
60. Hu, W., & Schatz, G. C., (2006). Theories of reactive scattering. The Journal of Chemical Physics, 125(13), 132301.
61. Hudson, E. R., (2009). Method for producing ultracold molecular ions. Physical Review A, 79(3), 032716.
62. Hummon, M. T., Yeo, M., Stuhl, B. K., Collopy, A. L., Xia, Y., & Ye, J., (2013). 2D magneto-optical trapping of diatomic molecules. Physical Review Letters, 110(14), 143001.
63. Hutzler, N. R., Lu, H. I., & Doyle, J. M., (2012). The buffer gas beam: An intense, cold, and slow source for atoms and molecules. Chemical Reviews, 112(9), 4803–4827.

64. Idziaszek, Z., & Julienne, P. S., (2010). Universal rate constants for reactive collisions of ultracold molecules. Physical Review Letters, 104(11), 113202.
65. Idziaszek, Z., Quéméner, G., Bohn, J. L., & Julienne, P. S., (2010). Simple quantum model of ultracold polar molecule collisions. Physical Review A, 82(2), 020703.
66. Jaksch, D., & Zoller, P., (2005). The cold atom Hubbard toolbox. Annals of Physics, 315(1), 52–79.
67. Jankunas, J., Bertsche, B., & Osterwalder, A., (2014). Study of the $Ne(^3P_2) + CH_3F$ electron-transfer reaction below 1 K. The Journal of Physical Chemistry A, 118(22), 3875–3879.
68. Jankunas, J., Bertsche, B., Jachymski, K., Hapka, M., & Osterwalder, A., (2014). Dynamics of gas-phase $Ne^*+ NH_3$ and $Ne^*+ ND_3$ Penning ionization at low temperatures. The Journal of Chemical Physics, 140(24), 244302.
69. Jankunas, J., Jachymski, K., Hapka, M., & Osterwalder, A., (2015). Observation of orbiting resonances in $He(^3S_1) + NH_3$ Penning ionization. The Journal of Chemical Physics, 142(16), 164305.
70. Jarvis, K. N., Devlin, J. A., Wall, T. E., Sauer, B. E., & Tarbutt, M. R., (2018). Blue-detuned magneto-optical trap. Physical Review Letters, 120(8), 083201.
71. Jochim, S., Bartenstein, M., Altmeyer, A., Hendl, G., Riedl, S., Chin, C., & Grimm, R., (2003). Bose-Einstein condensation of molecules. Science, 302(5653), 2101–2103.
72. Jones, K. M., Tiesinga, E., Lett, P. D., & Julienne, P. S., (2006). Ultracold photoassociation spectroscopy: Long-range molecules and atomic scattering. Reviews of Modern Physics, 78(2), 483.
73. Kim, J. B., Weichman, M. L., Sjolander, T. F., Neumark, D. M., Kłos, J., Alexander, M. H., & Manolopoulos, D. E., (2015). Spectroscopic observation of resonances in the $F + H_2$ reaction. Science, 349(6247), 510–513.
74. Kirste, M., Wang, X., Schewe, H. C., Meijer, G., Liu, K., Van, D. A. A., & Van De, M. S. Y., (2012). Quantum-state resolved bimolecular collisions of velocity-controlled OH with NO radicals. Science, 338(6110), 1060–1063.
75. Köhler, T., Góral, K., & Julienne, P. S., (2006). Production of cold molecules via magnetically tunable Feshbach resonances. Reviews of Modern Physics, 78(4), 1311.

76. Krems, R. V., & Dalgarno, A., (2004). Quantum-mechanical theory of atom-molecule and molecular collisions in a magnetic field: Spin depolarization. The Journal of Chemical Physics, 120(5), 2296–2307.
77. Krems, R. V., (2005). Molecules near absolute zero and external field control of atomic and molecular dynamics. International Reviews in Physical Chemistry, 24(1), 99–118.
78. Krems, R. V., (2006). Controlling collisions of ultracold atoms with dc electric fields. Physical Review Letters, 96(12), 123202.
79. Krems, R. V., (2008). Cold controlled chemistry. Physical Chemistry Chemical Physics, 10(28), 4079–4092.
80. Krems, R. V., Dalgarno, A., Balakrishnan, N., & Groenenboom, G. C., (2003). Spin-flipping transitions in 2 Σ molecules induced by collisions with structureless atoms. Physical Review A, 67(6), 060703.
81. Krzyzewski, S. P., Akin, T. G., Dizikes, J., Morrison, M. A., & Abraham, E. R. I., (2015). Observation of deeply bound Rb 2 85 vibrational levels using Feshbach optimized photoassociation. Physical Review A, 92(6), 062714.
82. Lara, M., Chefdeville, S., Hickson, K. M., Bergeat, A., Naulin, C., Launay, J. M., & Costes, M., (2012). Dynamics of the S (D 2 1)+ HD (j= 0) reaction at collision energies approaching the cold regime: A stringent test for theory. Physical Review Letters, 109(13), 133201.
83. Lavert-Ofir, E., Shagam, Y., Henson, A. B., Gersten, S., Kłos, J., Żuchowski, P. S., & Narevicius, E., (2014). Observation of the isotope effect in sub-kelvin reactions. Nature Chemistry, 6(4), 332–335.
84. Lee, Y. T., McDonald, J. D., LeBreton, P. R., & Herschbach, D. R., (1969). Molecular beam reactive scattering apparatus with electron bombardment detector. Review of Scientific Instruments, 40(11), 1402–1408.
85. Lemeshko, M., Krems, R. V., Doyle, J. M., & Kais, S., (2013). Manipulation of molecules with electromagnetic fields. Molecular Physics, 111(12, 13), 1648–1682.
86. Lique, F., Li, G., Werner, H. J., & Alexander, M. H., (2011). Communication: Non-adiabatic coupling and resonances in the F+ H2 reaction at low energies. The Journal of Chemical Physics, 134(23), 231101–231101.
87. Lu, H. I., Kozyryev, I., Hemmerling, B., Piskorski, J., & Doyle, J. M., (2014). Magnetic trapping of molecules via optical loading and magnetic slowing. Physical Review Letters, 112(11), 113006.

88. Lu, H. I., Rasmussen, J., Wright, M. J., Patterson, D., & Doyle, J. M., (2011). A cold and slow molecular beam. Physical Chemistry Chemical Physics, 13(42), 18986–18990.
89. Lukin, M. D., (2003). Colloquium: Trapping and manipulating photon states in atomic ensembles. Reviews of Modern Physics, 75(2), 457.
90. Masnou-Seeuws, F., & Pillet, P., (2001). Formation of ultracold molecules ($T \leq 200$ μK) via photoassociation in a gas of laser-cooled atoms. Advances in Atomic, Molecular, and Optical Physics, 47, 53–127.
91. McDonald, M., McGuyer, B. H., Apfelbeck, F., Lee, C. H., Majewska, I., Moszynski, R., & Zelevinsky, T., (2016). Photodissociation of ultracold diatomic strontium molecules with quantum state control. Nature, 535(7610), 122–126.
92. Micheli, A., Brennen, G. K., & Zoller, P., (2006). A toolbox for lattice-spin models with polar molecules. Nature Physics, 2(5), 341–347.
93. Mukherjee, N., Dong, W., & Zare, R. N., (2014). Coherent superposition of M-states in a single rovibrational level of H2 by Stark-induced adiabatic Raman passage. The Journal of Chemical Physics, 140(7), 074201.
94. Mukherjee, N., Dong, W., Harrison, J. A., & Zare, R. N., (2013). Communication: Transfer of more than half the population to a selected rovibrational state of H_2 by Stark-induced adiabatic Raman passage. The Journal of Chemical Physics, 138, 051101.
95. Mukherjee, R., (2019). Charge dynamics of a molecular ion immersed in a Rydberg-dressed atomic lattice gas. Physical Review A, 100(1), 013403.
96. Narevicius, E., & Raizen, M. G., (2012). Toward cold chemistry with magnetically decelerated supersonic beams. Chemical Reviews, 112(9), 4879–4889.
97. Neumark, D. M., (2005). Probing the transition state with negative ion photodetachment: Experiment and theory. Physical Chemistry Chemical Physics, 7(3), 433–442.
98. Neumark, D. M., Wodtke, A. M., Robinson, G. N., Hayden, C. C., & Lee, Y. T., (1985). Molecular beam studies of the $F + H_2$ reaction. The Journal of Chemical Physics, 82(7), 3045–3066.
99. Ni, K. K., Ospelkaus, S., De Miranda, M. H. G., Pe'Er, A., Neyenhuis, B., Zirbel, J. J., & Ye, J., (2008). A high phase-space-density gas of polar molecules. Science, 322(5899), 231–235.

100. Ni, K. K., Ospelkaus, S., Nesbitt, D. J., Ye, J., & Jin, D. S., (2009). A dipolar gas of ultracold molecules. Physical Chemistry Chemical Physics, 11(42), 9626–9639.
101. Nyman, G., & Yu, H. G., (2013). Quantum approaches to polyatomic reaction dynamics. International Reviews in Physical Chemistry, 32(1), 39–95.
102. Ospelkaus, S., Ni, K. K., Wang, D., De Miranda, M. H. G., Neyenhuis, B., Quéméner, G., & Ye, J., (2010). Quantum-state controlled chemical reactions of ultracold potassium-rubidium molecules. Science, 327(5967), 853–857.
103. Patterson, D., & Doyle, J. M., (2007). Bright, guided molecular beam with hydrodynamic enhancement. The Journal of Chemical Physics, 126(15), 154307.
104. Patterson, D., & Doyle, J. M., (2013). Sensitive chiral analysis via microwave three-wave mixing. Physical Review Letters, 111(2), 023008.
105. Patterson, D., & Doyle, J. M., (2015). A slow, continuous beam of cold benzonitrile. Physical Chemistry Chemical Physics, 17(7), 5372–5375.
106. Patterson, D., Schnell, M., & Doyle, J. M., (2013). Enantiomer-specific detection of chiral molecules via microwave spectroscopy. Nature, 497(7450), 475–477.
107. Pellegrini, P., Gacesa, M., & Côté, R., (2008). Giant formation rates of ultracold molecules via Feshbach-optimized photoassociation. Physical Review Letters, 101(5), 053201.
108. Prehn, A., Ibrügger, M., Glöckner, R., Rempe, G., & Zeppenfeld, M., (2016). Optoelectrical cooling of polar molecules to sub milli-Kelvin temperatures. Physical Review Letters, 116(6), 063005.
109. Quéméner, G., & Balakrishnan, N., (2009). Quantum calculations of H_2–H_2 collisions: From ultracold to thermal energies. The Journal of Chemical Physics, 130(11), 114303.
110. Quéméner, G., & Bohn, J. L., (2013). Ultracold molecular collisions in combined electric and magnetic fields. Physical Review A, 88(1), 012706.
111. Quemener, G., & Julienne, P. S., (2012). Ultracold molecules under control!. Chemical Reviews, 112(9), 4949–5011.
112. Quéméner, G., Balakrishnan, N., & Krems, R. V., (2008). Vibrational energy transfer in ultracold molecule-molecule collisions. Physical Review A, 77(3), 030704.

113. Regal, C. A., Greiner, M., & Jin, D. S., (2004). Observation of resonance condensation of fermionic atom pairs. Physical Review Letters, 92(4), 040403.
114. Regal, C. A., Ticknor, C., Bohn, J. L., & Jin, D. S., (2003). Creation of ultracold molecules from a fermi gas of atoms. Nature, 424(6944), 47–50.
115. Rellergert, W. G., Sullivan, S. T., Kotochigova, S., Petrov, A., Chen, K., Schowalter, S. J., & Hudson, E. R., (2011). Measurement of a large chemical reaction rate between ultracold closed-shell ca 40 atoms and open-shell yb + 174 ions held in a hybrid atom-ion trap. Physical Review Letters, 107(24), 243201.
116. Rellergert, W. G., Sullivan, S. T., Schowalter, S. J., Kotochigova, S., Chen, K., & Hudson, E. R., (2013). Evidence for sympathetic vibrational cooling of translationally cold molecules. Nature, 495(7442), 490–494.
117. Ruzic, B. P., Greene, C. H., & Bohn, J. L., (2013). Quantum defect theory for high-partial-wave cold collisions. Physical Review A, 87(3), 032706.
118. Sadeghpour, H. R., Bohn, J. L., Cavagnero, M. J., Esry, B. D., Fabrikant, I. I., Macek, J. H., & Rau, A. R. P., (2000). Collisions near threshold in atomic and molecular physics. Journal of Physics B: Atomic, Molecular and Optical Physics, 33(5), R93.
119. Santos, S. F. D., Balakrishnan, N., Lepp, S., Quéméner, G., Forrey, R. C., Hinde, R. J., & Stancil, P. C., (2011). Quantum dynamics of rovibrational transitions in H_2-H_2 collisions: Internal energy and rotational angular momentum conservation effects. The Journal of Chemical Physics, 134(21), 214303.
120. Sawyer, B. C., Lev, B. L., Hudson, E. R., Stuhl, B. K., Lara, M., Bohn, J. L., & Ye, J., (2007). Magneto electrostatic trapping of ground-state OH molecules. Physical Review Letters, 98(25), 253002.
121. Sawyer, B. C., Stuhl, B. K., Wang, D., Yeo, M., & Ye, J., (2008). Molecular beam collisions with a magnetically trapped target. Physical Review Letters, 101(20), 203203.
122. Sawyer, B. C., Stuhl, B. K., Yeo, M., Tscherbul, T. V., Hummon, M. T., Xia, Y., & Ye, J., (2011). Cold heteromolecular dipolar collisions. Physical Chemistry Chemical Physics, 13(42), 19059–19066.
123. Schnell, M., & Meijer, G., (2009). Cold molecules: Preparation, applications, and challenges. Angewandte Chemie International Edition, 48(33), 6010–6031.

124. Shagam, Y., & Narevicius, E., (2013). Sub-Kelvin collision temperatures in merged neutral beams by correlation in phase-space. The Journal of Physical Chemistry C, 117(43), 22454–22461.
125. Shagam, Y., & Narevicius, E., (2013). Sub-Kelvin collision temperatures in merged neutral beams by correlation in phase-space. The Journal of Physical Chemistry C, 117(43), 22454–22461.
126. Shuman, E. S., Barry, J. F., & DeMille, D., (2010). Laser cooling of a diatomic molecule. Nature, 467(7317), 820–823.
127. Shuman, E. S., Barry, J. F., Glenn, D. R., & DeMille, D., (2009). Radiative force from optical cycling on a diatomic molecule. Physical Review Letters, 103(22), 223001.
128. Simbotin, I., & Côté, R., (2015). Jost function description of near-threshold resonances for coupled-channel scattering. Chemical Physics, 462, 79–93.
129. Singh, V., Hardman, K. S., Tariq, N., Lu, M. J., Ellis, A., Morrison, M. J., & Weinstein, J. D., (2012). Chemical reactions of atomic lithium and molecular calcium monohydride at 1 K. Physical Review Letters, 108(20), 203201.
130. Smith, I. W., (2006). Reactions at very low temperatures: Gas kinetics at a new frontier. Angewandte Chemie International Edition, 45(18), 2842–2861.
131. Spaun, B., Changala, P. B., Patterson, D., Bjork, B. J., Heckl, O. H., Doyle, J. M., & Ye, J., (2016). Continuous probing of cold complex molecules with infrared frequency comb spectroscopy. Nature, 533(7604), 517–520.
132. Stark, K., & Werner, H. J., (1996). An accurate multireference configuration interaction calculation of the potential energy surface for the $F + H_2 \rightarrow HF + H$ reaction. The Journal of Chemical Physics, 104(17), 6515–6530.
133. Stoecklin, T., Voronin, A., & Rayez, J. C., (2002). Vibrational quenching of $N_2(v = 1, j_{rot} = j)$ by ^3He: Surface and close-coupling calculations at very low energy. Physical Review A, 66(4), 042703.
134. Strebel, M., Stienkemeier, F., & Mudrich, M., (2010). Improved setup for producing slow beams of cold molecules using a rotating nozzle. Physical Review A, 81(3), 033409.
135. Strecker, K. E., Partridge, G. B., & Hulet, R. G., (2003). Conversion of an atomic fermi gas to a long-lived molecular Bose gas. Physical Review Letters, 91(8), 080406.

136. Stuhl, B. K., Hummon, M. T., & Ye, J., (2014). Cold state-selected molecular collisions and reactions. Annual Review of Physical Chemistry, 65, 501–518.

137. Stuhl, B. K., Hummon, M. T., Yeo, M., Quéméner, G., Bohn, J. L., & Ye, J., (2012). Evaporative cooling of the dipolar hydroxyl radical. Nature, 492(7429), 396–400.

138. Suleimanov, Y. V., Tscherbul, T. V., & Krems, R. V., (2012). Efficient method for quantum calculations of molecule-molecule scattering properties in a magnetic field. The Journal of Chemical Physics, 137(2), 024103.

139. Tizniti, M., Le Picard, S. D., Lique, F., Berteloite, C., Canosa, A., Alexander, M. H., & Sims, I. R., (2014). The rate of the F + H_2 reaction at very low temperatures. Nature Chemistry, 6(2), 141.

140. Tscherbul, T. V., & Dalgarno, A., (2010). Quantum theory of molecular collisions in a magnetic field: Efficient calculations based on the total angular momentum representation. The Journal of Chemical Physics, 133(18), 184104.

141. Tscherbul, T. V., & Krems, R. V., (2008). Quantum theory of chemical reactions in the presence of electromagnetic fields. The Journal of Chemical Physics, 129(3), 034112.

142. Tscherbul, T. V., (2012). Total-angular-momentum representation for atom-molecule collisions in electric fields. Physical Review A, 85(5), 052710.

143. Ulmanis, J., Deiglmayr, J., Repp, M., Wester, R., & Weidemüller, M., (2012). Ultracold molecules formed by photoassociation: Heteronuclear dimers, inelastic collisions, and interactions with ultrashort laser pulses. Chemical Reviews, 112(9), 4890–4927.

144. Van De, M. S. Y., Bethlem, H. L., Vanhaecke, N., & Meijer, G., (2012). Manipulation and control of molecular beams. Chemical Reviews, 112(9), 4828–4878.

145. Van De, M. S. Y., Smeets, P. H., Vanhaecke, N., Jongma, R. T., & Meijer, G., (2005). Deceleration and electrostatic trapping of OH radicals. Physical Review Letters, 94(2), 023004.

146. Vogels, S. N., Onvlee, J., Chefdeville, S., Van, D. A. A., Groenenboom, G. C., & Van, D. M. S. Y., (2015). Imaging resonances in low-energy NO-He inelastic collisions. Science, 350(6262), 787–790.

147. Vogels, S. N., Onvlee, J., Von, Z. A., Groenenboom, G. C., Van, D. A. A., & Van De, M. S. Y., (2014). High-resolution imaging of velocity-controlled molecular collisions using counterpropagating beams. Physical Review Letters, 113(26), 263202.
148. Volpi, A., & Bohn, J. L., (2002). Magnetic-field effects in ultracold molecular collisions. Physical Review A, 65(5), 052712.
149. Wallis, A. O., & Hutson, J. M., (2009). Production of ultracold NH molecules by sympathetic cooling with Mg. Physical Review Letters, 103(18), 183201.
150. Wang, T. T., Heo, M. S., Rvachov, T. M., Cotta, D. A., & Ketterle, W., (2013). Deviation from universality in collisions of ultracold Li 2 6 molecules. Physical Review Letters, 110(17), 173203.
151. Weck, P. F., & Balakrishnan, N., (2006). Importance of long-range interactions in chemical reactions at cold and ultracold temperatures. International Reviews in Physical Chemistry, 25(3), 283–311.
152. Weinstein, J. D., DeCarvalho, R., Guillet, T., Friedrich, B., & Doyle, J. M., (1998). Magnetic trapping of calcium monohydride molecules at milli-Kelvin temperatures. Nature, 395(6698), 148–150.
153. Wigner, E. P., (1948). On the behavior of cross-sections near thresholds. Physical Review, 73(9), 1002.
154. Willitsch, S., (2012). Coulomb-crystallized molecular ions in traps: Methods, applications, prospects. International reviews in Physical Chemistry, 31(2), 175–199.
155. Willitsch, S., (2013). Ultracold ménage à trois. Nature Physics, 9(8), 461–462.
156. Willitsch, S., Bell, M. T., Gingell, A. D., Procter, S. R., & Softley, T. P., (2008). Cold reactive collisions between laser-cooled ions and velocity-selected neutral molecules. Physical Review Letters, 100(4), 043203.
157. Yang, B., Forrey, R. C., Stancil, P. C., Dos Santos, S. F., & Balakrishnan, N., (2012). Zero-energy resonances of hydrogen diatom isotopologs: Tuning quasi-resonant transitions in vibration space. Physical Review Letters, 109(23), 233201.
158. Zeppenfeld, M., Englert, B. G., Glöckner, R., Prehn, A., Mielenz, M., Sommer, C., & Rempe, G., (2012). Sisyphus cooling of electrically trapped polyatomic molecules. Nature, 491(7425), 570–573.

159. Zeppenfeld, M., Motsch, M., Pinkse, P. W., & Rempe, G., (2009). Optoelectrical cooling of polar molecules. Physical Review A, 80(4), 041401.

160. Zhelyazkova, V., Cournol, A., Wall, T. E., Matsushima, A., Hudson, J. J., Hinds, E. A., & Sauer, B. E., (2014). Laser cooling and slowing of CaF molecules. Physical Review A, 89(5), 053416.

161. Zhu, C., Krems, R., Dalgarno, A., & Balakrishnan, N., (2002). Chemistry of hydrogen fluoride in the interstellar medium. The Astrophysical Journal, 577(2), 795.

162. Zwierlein, M. W., Stan, C. A., Schunck, C. H., Raupach, S. M., Gupta, S., Hadzibabic, Z., & Ketterle, W., (2003). Observation of Bose-Einstein condensation of molecules. Physical Review Letters, 91(25), 250401.

Chapter 3

Near-Cold to Ultra-Cold Molecular Collisions

CONTENTS

3.1. INTRODUCTION

In the last two decades, the groundbreaking technologies of creating ultracold and cold molecules have offered cutting-edge experimentations for studying the essential event of collision physics (Werner et al., 2012). To a large degree, the current explosion of curiosity in molecular collisions has normally been generated by dramatic advancement of theoretical methods and experimental capabilities, which allow molecular collisions in order to be discovered in the quantum mechanical limit. Incredible experimental progress in this area has already been accomplished, and the authors, from the experimental viewpoint, give the review of studies for discovering the behavior of molecular collisions happening at temperatures varying from the K to the nK regime, along with for applications of creating ultracold molecules (Yang and Zhang, 2008).

Collision procedures controlled by the intra-particle interaction generally play an essential role in condensed matter physics, nuclear physics, molecular, and atomic physics, and chemistry. Comprehending the quantum nature of the collisions is significant for governing chemical reactions, enhancing energy efficiency, precision measurement, and exploring new phases of matter. The quantum collisions associated with reaction dynamics and chemical kinetics have been examined utilizing crossed-molecular beam collision and were stretched to the ultracold and cold regions with the laser-cooled atoms bringing Nobel prizes in physics and chemistry (McDonald et al., 1969; Herschbach 1987).

In current years, ultracold, and cold molecules, because of their intricacy of the internal level of energy comparing with the atoms, have provided exciting opportunities for accuracy measurements quantum computation many-body physics quantum chemistry and quantum simulation (Balakrishnan & Dalgarno, 2001).

To provide a brief introduction of this quickly emerging field, this chapter is motivated to assess current progress in cold, ultracold, and near-cold collisions relating to molecules from an experimental viewpoint (Hudson et al., 2006). To better classify the abundant collisional physics at the low temperature with delicate details, the technical terminology is adopted in $Temperature \geq 1K$ as near-cold, Temperature $\leq 1\mu$K as ultracold, and 1μK $<$ Temperature $< 1K$ as cold (Carr et al., 2009; Naulin & Costes, 2014). The researches of molecular collision can generally be dated to the 1970s when the molecular reactive collisions have usually been widely studied with the crossed-molecular beam methods with great accomplishment in attaining accurate information of the reaction dynamics for an atom-molecule system, like $F+H_2$ and $H+H_2$ systems (Harich et al., 2002). It is observed that, in a majority of such kinds of studies, the energy of the collision, i.e., comparative kinetic energy is in the region of the room temperature or much higher. At these temperatures, the molecules' de Broglie wavelengths are of the order of few thousandths of the nm (nanometer), and the collisions are normally understood as the collaboration of hard spheres traveling along the potential surfaces determined by electronic charge distribution(Micheli et al., 2006).

Even though some of the reaction scattering systems might be comprehended with the quantum-state resolution, the quantum state-resolved differential cross-sections are very difficult to acquire, since the collisional procedure is normally considered as the amalgamation of collisions of numerous partial waves which are rigorously overlapped.

In the meantime, the averaging over energies of collision, final, and initial states along with scattering angles, smears out and blurs any resonance pattern linked with the quantum effects. For the high-temperature collision, the influence of the single partial wave is bare to be determined experimentally (Weiner et al., 1999).

3.2. MOLECULAR COLLISIONS THEORY

Since this chapter analyzes the molecular collisions from the near-cold region to the ultracold region, where the de Broglie wavelengths of molecules for the translational motion become very large, and bearing in mind collision partners as the classical particles become worthless, focus would be laid on the quantum collision theory rather than classical collision theory, comprehensive treatment of which is provided in several textbooks (Bernstein, 1979). In the following, a detailed introduction will be given on the creation of a precise potential energy surface (PES) and the handling of collisions utilizing diverse scattering theories, triggering the complete understanding of the energy dependency of the collisional cross-section.

3.2.1. Potential Energy Surface (PES)

The PES plays a vital role in theories and concepts of molecular interactions and is the main ingredient in the prediction and computation of dynamics and structure. In Born-Oppenheimer approximation, there is a PES for every electronic state of a molecular system (Baranov, 2008; Eisert et al., 2015). The hindrance to effective scattering calculations, the case at all times, is the shortage of precise ab initio potentials for three systems comprising three or more than three atoms. Moreover, the inelastic and elastic collision cross-sections have been discovered sensitive to the particulars of interaction potential in numerous collisions by several theoretical models, because of the existence of quasi-bound state resonance. Usually, to obtain the accurate and high-precision PES around the region of curiosity, ab initio electronic structure calculations must be performed (Chalasinski & Szczesniak, 1994; Buchachenko et al., 2007).

To start, geometry modifications on the appropriate electronic states are executed, for instance, employing the restricted-spin coupled-cluster technique with double, single, and the non-iterative triple excitations or complete-active-space self-consistent-field (CASSCF) multi-reference internally contracted configuration interaction (MRCI) techniques (Werner & Knowles, 1985).

The basis sets like augmented correlation-consistent polarized valence triple-zeta (aug-cc-pVTZ) are utilized in such kinds of modifications together with frozen core calculation (Sachs et al., 1975; Dunning, 1989). Therefore, relative electronic energies and minimum-energy geometry of the particular electronic states are calculated.

Additionally, dissimilar basis sets can normally be utilized to examine the consequences of the basis set extension to the complete basis set (CBS) limit and the core electron association on the calculated quantities. Presently, MOLPRO programs have generally been popular for carrying out these calculations.

Then, ab initio functions of potential energy for the appropriate electronic states can normally be attained by fitting the polynomial in order to calculate the ab initio electronic energies (Jiang & Guo, 2013; Hollebeek et al., 1999).

Various methods can be utilized to give reasonably precise fits to the ab initio data, like rotating morse spline (RMOS), the reproducing kernel Hilbert-space (RKHS) interpolation technique, the distributed approximating functional technique many-body expansion method, and the generalized London-Eyring Polanyi-Sato double-polynomial (GLDP) approach (Garcia & Lagana, 1985; Sorbie & Murrell, 1975).

These fitting techniques should satisfy three criteria: the fitting process must be simple, the consequential surface must be smooth and consent with the data of ab initio, without oscillations amongst the ab initio points, and its mathematical form must be realistic physically to reduce the required number of ab initio geometries in order to obtain the surface with correct topology and features.

In the calculation above, the reservations can come from several sources, which are the ab initio theory level used to roughly solve Schrodinger's equation, the basis set's size engaged in the calculations, number of the grid points utilized, the dimensionality of the potential energy source, and the precision of analytic fit to the calculated energies (Ho & Rabitz, 2000). Since the precise potential energy source is the first constituent for a precise study of scattering, it is thus very essential to control the doubts from several sources to attain optimum accuracy of the potential energy source.

The calculation of potential energy sources for NH−NH might be exemplary. The potential energy source can be utilized to conclude the cross-sections for spin-changing and elastic NH−NH collisions, which determines the possibility for the process of evaporative cooling.

Guillaume S. F. Dhont et al. made 4-D ab initio potential energy sources for the NH−NH complex, which associate with two distinct NH molecules in the $^3\Sigma^-$ electronic ground state (Dhont et al., 2005).

The PES of quintet state (S=2) was computed at coupled cluster level with the RCCSD(T) method. The singlet and triplet states defined with multi-reference functions of wave and the exchange splittings amongst the potential surfaces were computed with CASSCF techniques followed by third or second-order perturbation theory.

The full PESs were computed as the function of 4 intermolecular Jacobi coordinates, having the aug-cc-pVTZ basis on H and N atoms (Liu et al., 2014).

The analytical demonstration of the potentials was provided by expanding their dependency on molecular orientations in the coupled spherical harmonics and signifying the dependency of the coefficients of expansion on the intermolecular distance by the RKHS technique. A following study offered novel RCCSD(T) calculations, which amended the inaccurate behavior in the long range because of the lack of size steadiness in the open-shell RCCSD(T) technique (Janssen et al., 2009).

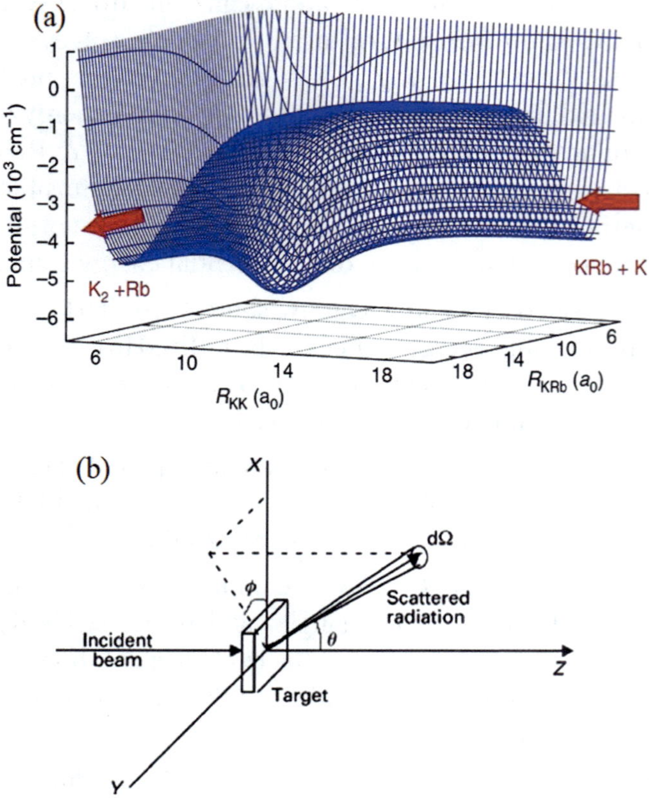

Figure 3.1 (a) K-KRb PES (potential energy surface) for collinear geometry. The 2-D cut through the energetically lowermost 2A′ adiabatic PES of KRbK trimer as the function of K-K and K-Rb bond lengths beside the collinear geometry. (b) Standard scattering geometry for measuring differential scattering cross-section

Source: https://www.nature.com/articles/ncomms15897.

3.2.2. Scattering Theory

Quantum scattering computations of molecular and atomic collisions, particularly atom-diatom chemical reactions, have traditionally been carried out using time-dependent and time-independent quantum-mechanical methods. The time-dependent wave packet approach is popular for managing molecular collisions at room temperature and higher (Althorpe & Clary, 2003). Though, it becomes tough to execute at very low energies of collision, since the time advancement of the wave packet is quite slow, and spreading it until this wave packet reaches the asymptotic regime takes a quite long time. Moreover, it is problematic to converge wave packet computations at very low energies of scattering, and it inclines to trigger unphysical reflections. Therefore, time-independent approaches are preferred to manage the low-energy collisions.

In high-energy physics experiments, a beam of particles hits a target composed of particles. By detecting the by-products, one aims to study the interactions that occur during the collision.

Collisions can be rather intricate. For example, the particles involved may be not elementary (protons) or they may be elementary (electrons and positrons):

$$\mathrm{p} + \mathrm{p} \rightarrow \mathrm{p} + \mathrm{p} + \pi^0$$

$$\mathrm{p} + \mathrm{p} \rightarrow \mathrm{p} + \mathrm{n} + \pi^+$$

$$\mathrm{e}^+ + \mathrm{e}^- \rightarrow \mu^+ + \mu^-$$

The final products may not be the same as the initial particles. Some particles may be created. Here collisions are like reactions in which a + b → c + d + e +. . . We have scattering when the particles in the initial and final state are the same a + b → a + b.

3.3. NEAR-COLD COLLISION EXPERIMENT

In the thick molecular clouds of the interstellar medium, which are generally at the base of star development, gas-phase chemistry is categorized by low temperatures and ultralow gas density, normally the temperature of cold cores might reach 5–10 K (Herbst & Yates, 2013). The Boomerang Nebula has a temperature of 1 Kelvin (Sahai & Nyman, 1997), falling quite below the temperature of cosmic background radiation Temperature=2.8K. Usually, gas-phase reactions comprise two neutral species that have activation energy barriers and the coefficients k of reaction rate are given by the following Arrhenius expression:

$$k(T) = A(T)\exp(-E_a/T) \quad (1)$$

where; T →temperature; E_a →activation energy in K; A(T) →weakly temperature-dependent term is called the pre-exponential factor.

However, radical-radical, and atom-radical processes take place quickly with $E_a = 0$ and the pre-exponential factor reaches the collision limit.

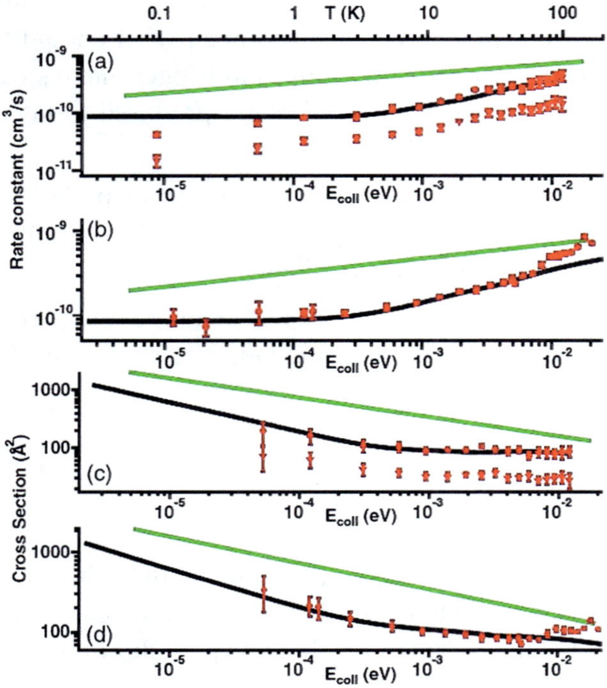

Figure 3.2. Rate constants for (i) the Ne* + NH_3 reaction (ii) the Ne* + ND_3 reaction. Panels (iii) and (iv) display the cross-sections for Ne*+ND_3 and Ne* + NH_3, correspondingly.

Source: https://aip.scitation.org/doi/abs/10.1063/1.4883517.

Note: Red symbols give the experimental data, whereas black lines are the outcomes from the MQDT computations. The green lines display the rate anticipated based on the Langevin model having unit reaction possibility at short range.

It has frequently been supposed the reaction rate coefficient k for the reactions can be estimated by hard-sphere relation below:

$$k(T) = A(T) = k(300K)(T/300)^{0.5} \quad (2)$$

where usual value for k(300K) is around 1.0–3.0×$10^{-11}cm^3/s$. With this supposition, the reaction rate coefficients are significantly small. However, numerous crossed-beam experimentations with variable low energies of collision in combination with the ab initio quantum chemical computations have displayed that neutral-neutral reactions have 0 threshold energy and advance through intensely bound energized complexes, which generally have drastically changed the understanding of low temperature (T) interstellar chemistry over the past two decades (Herschbach, 2009). Several reactions have been discovered to have slight temperature dependency or become quicker with the decrease in temperature, and their reaction rate coefficients can frequently be fit to the feeble inverse power law in T. Even though the T is still high as compared to the cold region and the collision energy is far above the pure s-wave scattering, the near-cold region will open the limited number of partial waves through which the resonance signatures aren't entirely blurred by the partial wave averaging.

3.3.1. Crossed Molecular Beam

Since the discovery of crossed molecular beam method, it has generally become the typical tool for studying inelastic collisions and the reactions in molecule-molecule and atom-molecule systems (Schnieder et al., 1995; Xiao et al., 2011). Attractive agreement between experiment and theory in the benchmark 3-atom $F + H_2 \rightarrow HF + H$ and $F + HD \rightarrow HF + D$ reactions has been accomplished, where scattering resonance characteristics have been completely rationalized(Qiu et al., 2006). The agreement has been attained in high-resolution crossed beam experiments for ideal 6-atom $Cl + CHD_3 \rightarrow HCl + CD_3$ and 4-atom $HD + OH \rightarrow H_2O + D$ systems (Wang et al., 2011; Vogels & Karman, 2018). However, for several years the research has been carried out at high-temperature regions with contributions from numerous partial waves and the energy of collision large enough to overcome the traditional energy barriers.

Currently, encouraged by curiosity in the collision procedures in astrophysical environments, substantial advancement has been made in the crossed molecular beam research of inelastic collisions and chemical reactions. In the crossed molecular beam experimentations, collision partner A having velocity v_1 and partner B having velocity v_2 in the laboratory frame will possess the collision energy:

$$E_c = 1/2\mu v_r^2 = 1/2\mu(v_1^2 + v_2^2 - 2v_1v_2 cos\chi),$$

where; μ →reduced mass; v_r →relative velocity; χ →beam intersection angle.

The least comparative collision energy is given as $E_c = \mu v_1^2(1 - cos\chi)$ when $v_1 = v_2$. Thus, both small relative velocity and small beam intersection angle are needed to acquire low energy of collision in the crossed-molecular beam experimentations.

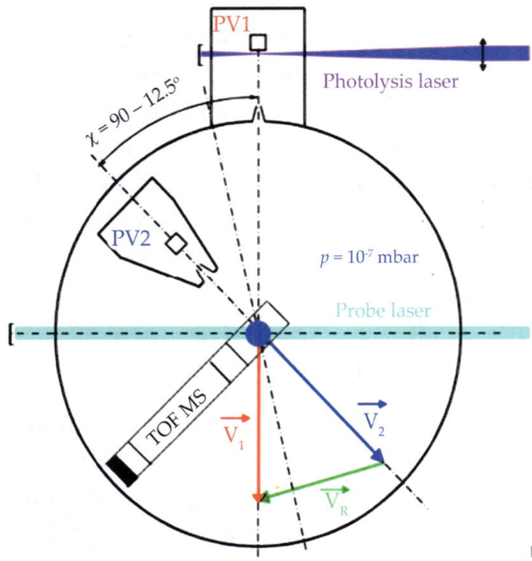

Figure 3.3. Experimental setup of crossed molecular beam for the study of the reaction dynamics having a tunable angle.

Source: https://www.tandfonline.com/doi/abs/10.1080/0144235X.2014.957565.

It has been decreased to nearly 4 K as can be signified by Figure 3.3 (Berteloite et al., 2010; Bergeat et al., 2015). The experiments have established witnessing of quantum effects in the molecular collisions, like orbiting resonance and the interferences arriving at the near-cold regime (Vogels et al., 2018; Liu & Liu & Luo, 2021).

Chef Deville et al. stated quantum mechanical and experimental studies on CO (j=0) +H_2 (j=0) → CO (j=1) + H_2 (j=0) inelastic collisions. In his experiment, the beams were generally from the cryogenically chilled Even Lavie fast-pulsed spigots, which permit not only perfect control over the distribution of velocity but also appease a majority of the inner state population to ground rotational state. Beam intersection angle was reduced to 12.5 degrees. The integral cross-section was computed in the threshold regime of CO (j=0→ j—1) cm^{-1} transition between 23 and 3 cm^{-1}, conforming to 33.09 and 4.32 K, which exhibit resonance structures. This specified the stricter requirement on the PES for fully defining the experimental measurements (Vogels et al., 2018).

In the following study of $S(^1D_2)$+HD(j–0) reaction at the energies from 54.2 milli eV down to 0.46 milli eV clear measurements of the integral cross-sections for the DS + H and HS + D products and their ratios of branching were compared to the adiabatic time-independent quantum mechanical computations (Lara et al., 2012). The product channel ratio of branching ratio and the low energy resonance doesn't feature between 1.5 and 5 milli eV in HS + D channel can generally be reproduced by the theoretical computations. This inconsistency was accredited to the limitation of the single adiabatic electronic PES rather than the complete time-dependent quantum mechanical management, bearing in mind all 4 PES and their non-adiabatic couplings for scattering computations. Presently for O_2–H_2 inelastic collisions carried out with the similar experimental apparatus with the complete quantum close-coupling scattering computations by utilizing the 4-D ab initio PES attained by coupled-cluster approach whereas using double and single excitation having perturbative contributions from associated triple excitations and the large set of atomic basis orbitals, entire characterization of completely determined partial wave resonances and outstanding settlement with the experimental integral cross-sections has been attained in 3.7–20 cm^{-1} range of energy, as demonstrated in Figure 3. 4 (Chefdeville et al., 2013).

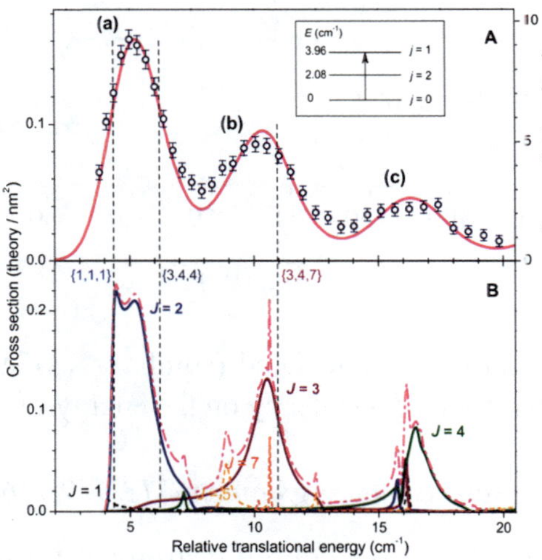

Figure 3.4. Collisional energy dependency of integral cross-sections for O_2 excitation |j = 0i, N=1 → |j = 1i,N=1. (i) Data of experiment with para-H_2. (B) Theoretical outcomes with partial waves J = 4, 3, 2 (solid-lines), partial waves J = 5–7 and J=1 (dashed-lines), integral cross-section (dashed-dotted line).

Source: https://science.sciencemag.org/content/341/6150/1094.

In the first scattering experiment with Stark decelerated molecules, the pack of OH radicals was decelerated and the velocity of the pack was adjusted between 33 and 700 ms^{-1}, then scattered with the pulsed beam of Xenon in the crossed-beam geometry having the 90-degree intersection angle. The state to state fundamental scattering cross-sections was measured and then calculated as the function of collision energy between 50 and 400 cm^{-1} with an energy resolution of nearly 13 cm^{-1}. Figure 3.5 displays the resulting comparison of collision energy dependency of the calculated and measured comparative cross-sections.

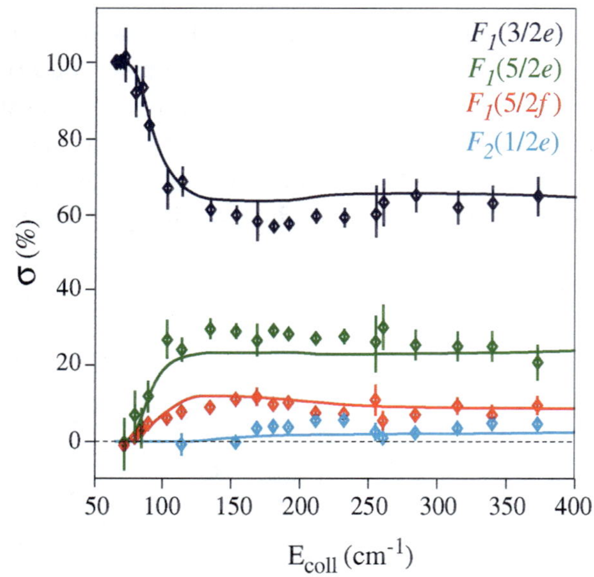

Figure 3.5. Comparison of the calculated (solid curves) and measured (data points) relative cross-sections dependency on the energy of the collision.

Source: https://science.sciencemag.org/content/313/5793/1617.

Besides, this method is also a helpful instrument for imaging scattering resonances in the atom-molecule collisions. Utilizing the amalgamation of velocity mapped imaging and Stark deceleration technique, the impressions of the scattering resonances in DCSs (differential cross-sections) for inelastic collisions amongst He atoms and NO radicals are perceived (Vogels et al., 2015). In the experiment, the slow Helium beams from the cryogenically chilled Even-Lavie valve overlapped with a Stark-decelerated beam of NO radicals at the angle of 45 degrees, consequential in the energies of collision as low as around 12.5 cm^{-1}, corresponding to the temperature of about 12.0 K. The comparative velocity vector was select almost perpendicular to the Helium direction of propagation, which occasioned in the optimum energy resolution of nearly 0.3 cm^{-1} at the collision energies close to the resonances. With the experimental arrangements, the images of scattering were measured at the energies of collision between 13 & 19 cm^{-1}, which generally has been anticipated as where the scattering resonances occur by the high-level mechanical scattering computations. As displayed in Figure 3.6, striking discrepancies in an angular distribution were perceived when changing the energy in some small stages over the resonances. As perceived from Figure 3.6, the diameter of images for a $|J = 1/2, fi \rightarrow |J = 5/2, fi$ scattering channel is reaching 0 when the energy of collision approaches the thermodynamic verge of a channel, which discloses the apparent threshold behavior. Generally, the measured differential cross-sections agree quite well with that attained from ab initio computations. As the scattering resonance is sensitive to the PES, this might be the promising approach to provide essential information for the hypothetical models (Chefdeville et al., 2012; Vogels et al., 2015).

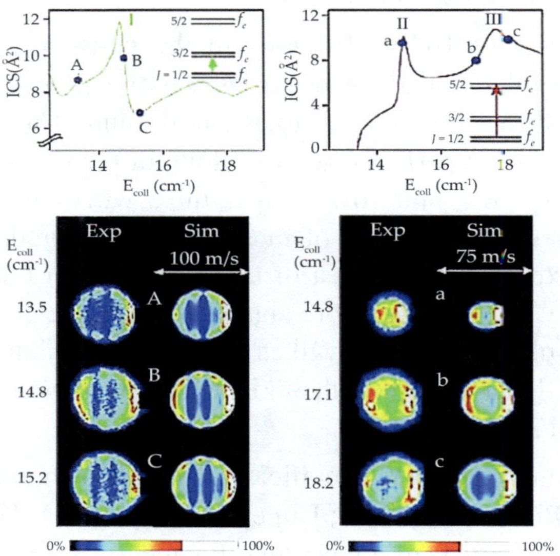

Figure 3.6. Simulated and experimental ion images at chosen collision energies for the channels of |J = 1/2,fi → |J =5/2,fi (right) and |J = 1/2,fi → |J = 3/2,ei (left) in the inelastic collisions amongst NO and Helium.

Source: https://science.sciencemag.org/content/350/6262/787.

3.3.2. Intra-Beam Collision

3.3.2.1. CRESU

Some types of collision studies can generally be regarded as an intra-beam collision, where the collisions between diverse collision partners can usually be studied inside a single molecular beam. The first among these is the Cinetique de Reaction en Ecoulement Supersonique Uniforme (CRESU). In a CRESU apparatus, the continuous, unvarying supersonic flow of persistent temperature and density is produced by the gas expansion from the reservoir at reasonable pressure through the convergent-divergent Laval nozzle (Sowokinos, 2001). The schematic representation is displayed in Figure 3.7. At the Laval nozzle exit, as there isn't any additional expansion downstream, the parameters of flow (i.e., density, temperature, velocity, and pressure) don't display any radial and axial discrepancies in the center of a jet. The velocity of diffusion is always insignificant concerning bulk velocity, thus averting the problem of condensation with the utilization of cryogenically chilled cells. As an outcome, in such kinds of expansions, severely supersaturated conditions exist.

Usual gas density 10^{16} to 10^{17} cm^{-3} in the constant supersonic flow confirms that regular collisions occur during the process of expansion and succeeding flow, sustaining thermal equilibrium. Thus, such constant supersonic flows give a perfect reactor of flow for the research of chemical reactivity at very low and low temperature, where these kinematics, most apparently the coefficients of rate can be measured. Observe, in a crossed beam experiment, the idea of temperature isn't effective, leading to inhomogeneous supersonic, more significantly, the scale of time for the reaction to take place doesn't prevail as the single collision event controls. Therefore, only dynamics rather than kinematics prevails in crossed-beam studies (James et al., 1998).

The 1st measurements of the coefficients of rate down to 20 Kelvin were stated in 1984 (Rowe et al., 1984; Dupeyrat et al., 1985). The apparatus of CRESU provides the versatile technique for measuring the rate constants and generally has been made practical to several neutral molecule-molecule and atom-molecule reactions comprising radical kinds in the range of 1–100K (Herbst, 1995). The least temperature in the CRESU experiment has accomplished 7 Kelvin so far with the pre-cooled reservoir by liquid nitrogen. In spite of success, it does have few drawbacks: supersonic flow is constant for just tens of cms, which limits the offered hydrodynamic time to few hundreds of microseconds, and the new nozzle for every set of conditions, condition of the large pumping capabilities, along with a restricted range of entire gas densities (Dubois et al., 2016).

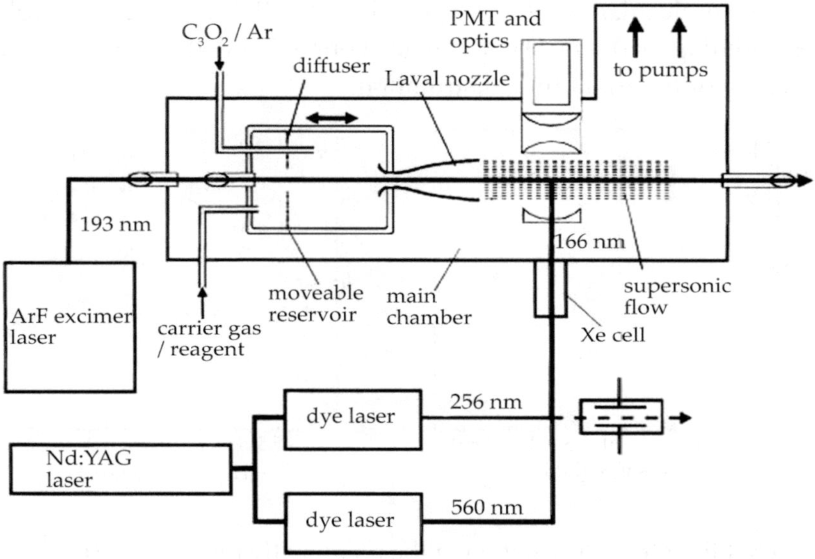

Figure 3.7. Schematic representation of the apparatus of CRESU for the learning of C(3P) reaction kinetics. The atoms of C(3P) were made by photolysis of the C_3O_2 at 193nanometers and perceived by laser tempted fluorescence in a vacuum ultraviolet.

Source: https://pubs.rsc.org/en/content/articlehtml/1999/cp/a900449a.

3.3.2.2. SARP

The 2nd one is co-expanding gases in the assorted molecular beam succeeding stark-induced adiabatic Raman passage (SARP) (Perreault et al., 2017). This approach prepares both of the colliding partners in the single co-expanding supersonic beam, therefore brings down the collision temperature to 1 K, limiting scattering to p and s partial waves. By transferring coherently the population of 1 partner to the excited state and configuration of a molecular bond axis perpendicular/parallel to the comparative velocity of colliding partners utilizing SARP, the HD molecules are made in particular *m* states of the rovibrational energy eigenstate. Therefore, complete stereodynamic control of a scattering event can normally be accomplished. In Figure 3.8, the flight time spectrum of the scattered HD ($j = 1$, $v = 1$) establishes strangely stereodynamic effects. When a molecular-bond axis is lined up perpendicular to the relative velocity of colliding partners, the flight time distribution of HD ($j = 1$, $v = 1$) is approximately three times strong matched to when the molecular-bond axis is lined up parallel to the relative velocity of the colliding partners. This can be comprehended by a partial-wave examination of the flight time distribution.

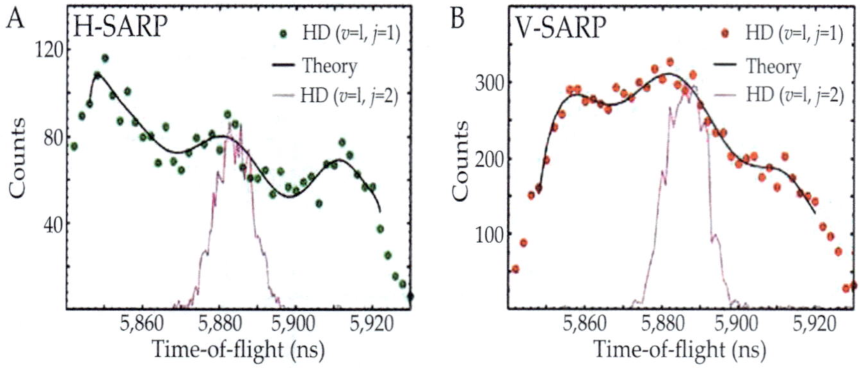

Figure 3.8. Flight time distribution of the rotationally relaxed HD (j = 1, v = 1).

Source: https://science.sciencemag.org/content/358/6361/356.

Angular distributions can usually be taken out from time-of-flight distributions measured experimentally of the scattered HD molecules. Figure 3.9 exhibits the angular distribution of a scattered HD ($j = 0$, $v = 1$) for the HD molecular-bond axis aligned perpendicular and parallel to the relative velocity between D_2 and HD. Thus the complete stereodynamics in the molecular collisions is studied for HD–D_2 and HD–H_2 systems utilizing measured partial wave analysis and 4-vector correlations (Barnwell et al., 1983).

3.3.2.3. Cryofuge

The 3rd approach is cryofuge which utilizes centrifugal force in order to bring cryogenically chilled molecules to the kinetic energies below one Kelvin in a laboratory frame (Wu et al., 2017). In this experiment, apparent collisional losses are detected because of acquired density higher than $1x10^9$/ cm and the times of interaction longer than 25 milliseconds for the samples of CH_3F and ND_3.

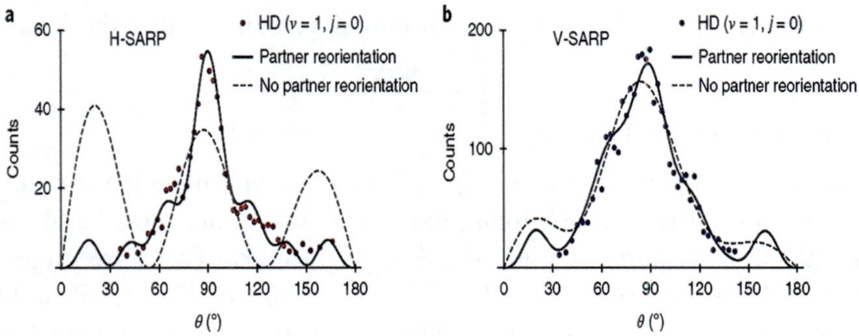

Figure 3.9. Angular distribution of a scattered HD (j = 0, v = 1) derived from the flight time measurements of REMPI-produced ions.

Source: https://www.nature.com/articles/s41557-018-0028-5.

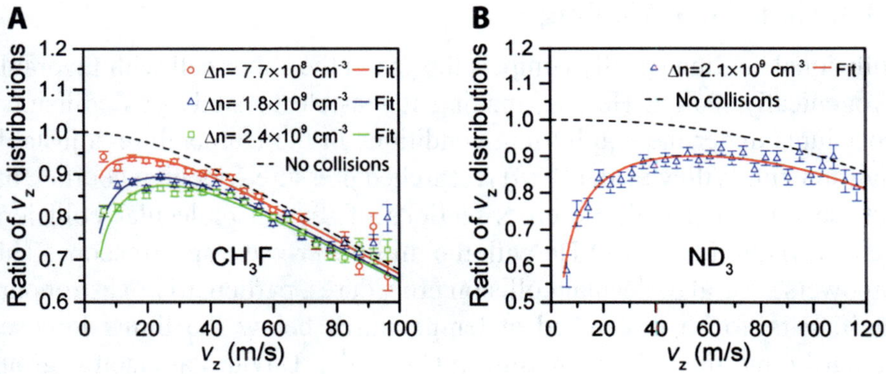

Figure 3.10. Depletion from the molecular collisions in a centrifuge-decelerated beam for (i) CH3F and (ii) ND_3.

Source: https://science.sciencemag.org/content/358/6363/645.

In Figure 3.10, the collisional loss dependency on the longitudinal velocity v_z and molecular density n of a guided beam is inspected. In the guided beam, the loss of collision alters the longitudinal velocity v_z distribution according to the equation:

$$P = P_0^{\left(\frac{-k_{loss} nL}{v_z}\right)}$$ where $k_{loss} = \sigma_{loss} v_{rel} \rightarrow$ loss rate coefficient.

Therefore, the ratio for two longitudinal velocity v_z distributions can normally be explained by the $\ddot{u}\left(\frac{-\alpha \pm_{th} \Delta}{v_z}\right)$ model,

where k_{th} →theoretically anticipated loss rate coefficient; $\Delta n = n_{high} - n_{low}$ → density difference. Utilizing the Langevin capture model and semi-classical Eikonal approximation, the theoretical rates of loss are projected to be $k_{th}^{ND}3 = 1.3 \times 10^{-9} cm^3/s$ and $k_{th}^{CH3F} = 7.7 \times 10^{-10} cm^3/s$ at 1.1 K and 0.8 K collision energy, correspondingly. The values are approximately 60% and 40% larger than the theory, which discloses the inaccuracy of a Langevin model. In addition, the loss rate coefficient ratio agrees with the semi-classical computations which forecasts the rate coefficient $k \propto d^2$, and d being the dipole moment of a molecule (Liu & Luo, 2021).

3.3.3. Collisional Cooling

Collisional cooling usually compels the gas in the vapor cell with favorably cryogenically chilled He gas, making it feasible to study gas collisional procedures under the equilibrium conditions in the temperature-adjustable environment. Willey et al. (1988) researched pressure widening coefficients and the rotational inelastic cross-sections of several molecular collisions such as CH_3F-He and CO-He with the millimeter-wave spectroscopy. This empowers several molecular collision procedures, particularly of astronomy significance to be researched at temperatures below the limits imposed ordinarily by the vapor pressures. Currently, David Patterson's group examined the conformational dynamics of the large polyatomic molecules (Drayna et al., 2016). In their experiment, conformational relaxation of the 1,2-propanediol in collision with He in a 6 Kelvin cryogenic environment is noticed directly utilizing microwave spectroscopy.

3.3.4. Target Collision

Early studies of target collision were made directly after the ability to ensnare molecular samples was comprehended. An instance of the first measurement is the OH−N_2 complete scattering cross-section at ambient temperature 295 Kelvin in the magnetic trap, where the molecules of OH are loaded from the Stark decelerator (Sawyer et al., 2007). This is currently followed by the collision learning of the magnetically stuck CH_3 with surrounding H_2 molecules (Liu et al., 2017).

To reach the lower energies of collision, Sawyer et al. measured the OH–D_2 and OH-He cross-sections by utilizing the liquid-nitrogen chilled supersonic beam of D_2 or He to attack the target (Sawyer et al., 2008). The experiment discloses the striking characteristic in OH–D_2 collisions, the pronounced peak in cross-section at 305 cm^{-1} appears, which recommends resonant transfer of energy probably backing from $J = 1 \rightarrow J = 3$ transition or the collision-induced decay of a ${}^1\Sigma_g^+ D_2$ molecule. After this, the energy of the collision is lowered further utilizing the slow beam of ND_3, which was the cooled buffer gas and consequently velocity filtered by the 90^0 bent hexapole guide, striking with the magnetically trapped molecules of OH (Sawyer et al., 2011). The assembly of this experiment is represented in Figure 3. 11.

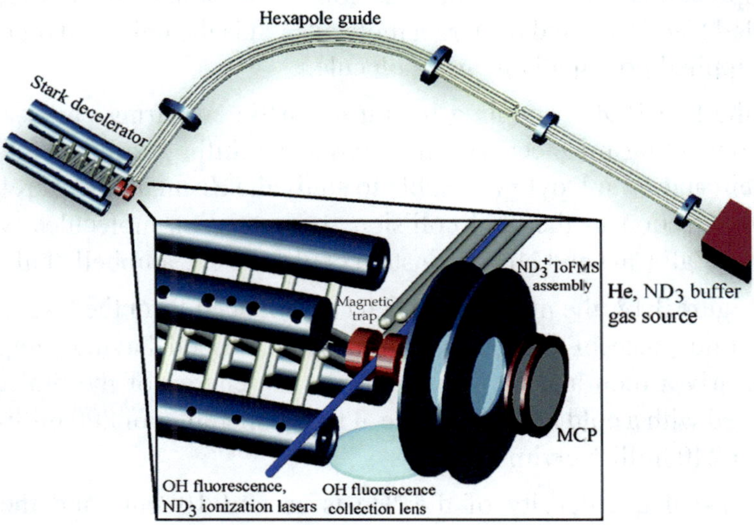

Figure 3.11. Representation of the joined experimental assembly for the learning of low-energy collision of OH and ND_3.

Source: https://pubs.rsc.org/en/content/articlelanding/2011/cp/c1cp21203f.

3.4. COLD COLLISION EXPERIMENT

In the group of optical, molecular, and atomic physics, astrochemistry, and chemical physics, there exists the great desire of learning about cold molecular collisions, since they are perfect to perceive the quantum nature of the molecular collisions as an angular momentum becomes strictly constrained, ultimately approaching the limit l = zero where only the s-wave scattering can occur and de Broglie wavelength linked with the relative velocity becomes analogous to or greater than the size of the reactant (Thomas et al., 2004). It is the perfect platform to attain accurate information of the inter-atomic interaction perspective and to study cold chemistry. Furthermore, it can be utilized to test several ab initio computations on the molecular PES, and dissimilar scattering theories in attaining differential and integral cross-section.

3.4.1. Buffer Gas Cooling

This approach is dependent on elastic collisions amongst the cryogenically precooled He/Ne gas and the target molecules. It is the universal method and can be applied principally to any molecule.

In the late 1990s, coupling in-situ magnetic ensnaring empowered by super-conducting magnets with the cryogenic chilling, group of Jonathan Weinstein and John Doyle were able to study the Zeeman rate of relaxation and cross-section in the cold collisions of numerous molecules with He, including CaF CaH and NH (Weinstein et al., 1998; Campbell et al., 2009).

In Figure 3.12, the molecule of CaH is produced from the laser ablation of the solid piece of CaH_2 within the cryogenic cell having temperature up to nearly a thousand K. Consecutively, the energy of the molecules is exchanged with a cold bath of the ^{4}He at the temperature of 800 milli-Kelvin or ^{3}He at 240 milli-Kelvin.

The usual gas density of the ^{4}He is around 10^{16}cm^{-3} and the cross-section for the elastic scattering is supposed to be of the order of nearly 10^{-14}cm^2. After nearly a hundred collisions, the molecules are chilled to the similar temperature of the buffer gas and consequently be trapped by the superconducting magnetic field placed outside a cell and settled in the anti-Helmoltz arrangement, whose strength is around 3 T, conforming to the trap depth of nearly 2 K for the molecules of CaH (Maussang et al., 2005). The molecules are normally excited via $|B^2\Sigma^+,v' = 0$i $\leftarrow$ $|X^2\Sigma^+,v'' = 0$i transition and observed utilizing rotational transition $|N' = 0,J' = 3/2$i $\leftarrow$ $|N'' = 0,J'' = 1/2$i from the radiative decay of $|B^2\Sigma^+, v' = 0$i $\rightarrow$ $|X^2\Sigma^+, v'' = 1$i.

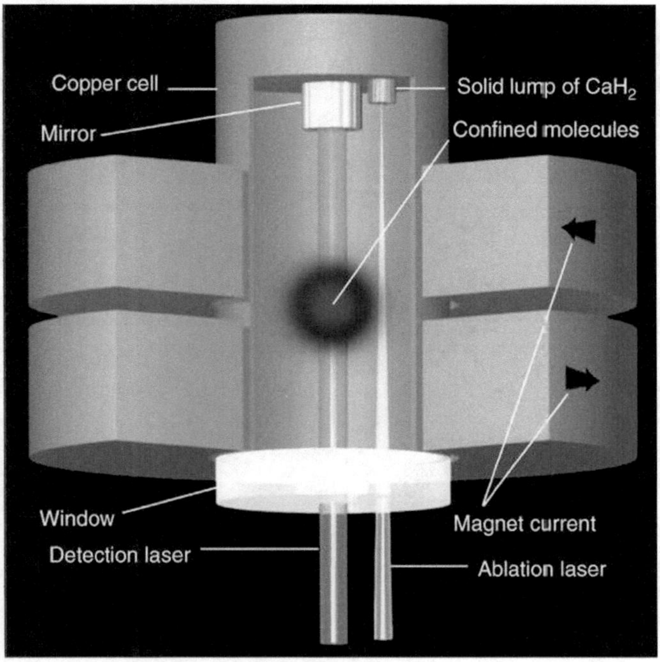

Figure 3.12. Diagram of magnetic ensnaring of the molecules of CaH collisionally cooled by Helium, which in term cryogenically chilled by the dilution refrigerator.

Source: https://www.nature.com/articles/25949.

By noticing the time development of these molecules, the dynamics of the trapping and loading procedure can be studied.

Excluding the in-situ loading by laser ablation, the molecules can be loaded utilizing the molecular beam (Campbell et al., 2007). As displayed in Figure 3.13, the NH beam, which is transformed from ammonia by the room temperature glow discharge, arrives at the cryogenic cell via an orifice. The cold N-NH collisions at a temperature of nearly 600 milli-Kelvin are also comprehended (Hummon & Tscherbul, 2011). The N + NH snare loss rate coefficient accrediting to the inelastic relaxation is computed, and the elastic to inelastic collisions ratio is large over the temperature range of 10 mK to 1 K from the ab initio quantum scattering computations.

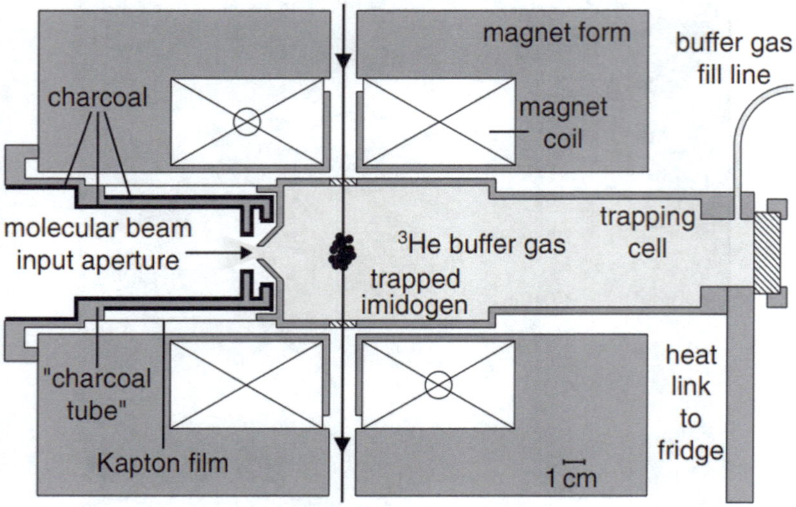

Figure 3.13. Schematic representation loading beam of NH into a magnetic trap.

Source: https://journals.aps.org/prl/abstract/10.1103/PhysRevLett.98.213001.

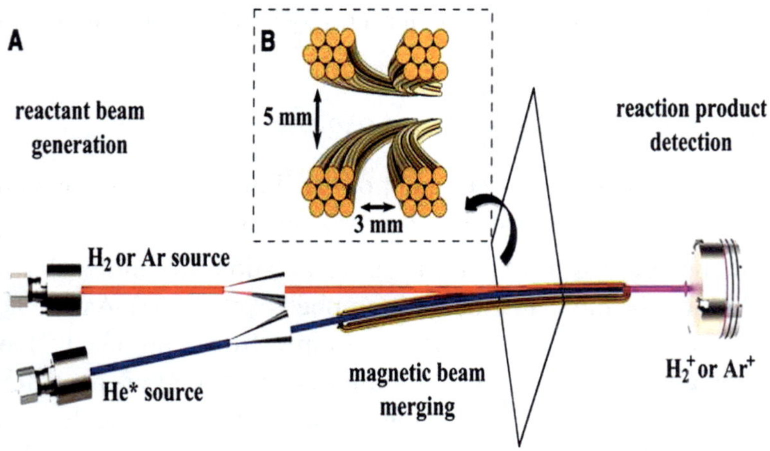

Figure 3.14. Diagram of the setup of Penning ionization reaction with a merged molecular and atomic beam. The beam of metastable He (blue) moving beside a curved magnetic quadrupole guide combines with the other beam of Ar (red) or H_2.

Source: https://science.sciencemag.org/content/338/6104/234.

3.4.2. Merged Molecular Beam

As seen in the former section, the relative energy of the collision is in the near-cold regime or high-temperature region in usual crossed molecular beam experimentations. To decrease the relative energy of collision to a cold region, diverse proposals utilizing the supersonic expansion method have been developed. Qi Wei et al. suggested integrating slowed beam from the counter-rotating nozzle with the pulsed molecular beam in a way that the angle of scattering approaches 0, and the relative velocity can smoothly be tuned to the ultracold collision region by varying the repetition frequency of a counter-rotating nozzle and condition of the supersonic expansion for the pulsed molecular beam (Wei et al., 2012).

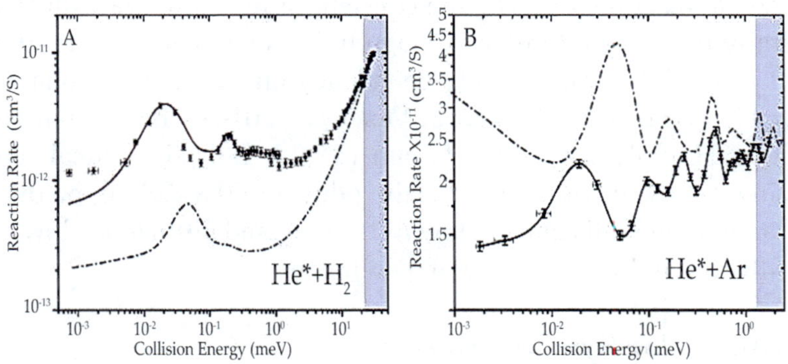

Figure 3.15. Measurement of the reaction rate for (i) (3S)He* & H_2, (ii) (3S) He* & Ar. In panel A, the red solid-line is the computed reaction rate by utilizing the Tang-Toennies potential. In panel B, the black-solid line is the guide to an eye.

Source: https://science.sciencemag.org/content/338/6104/234.

This concept of merging beam has been realized first in the usual He* + $H_2 \rightarrow$ He + H_2^+ Penning experiment of ionization where metastable He produced by the dielectric barrier discharge is directed by the curved magnetic quadrupole guide having large curvature and the small angle, and join with the pulsed beam of the molecular hydrogen at 0 degrees (Henson et al., 2012).

The diagram is displayed in Figure 3.14. The relative velocity can normally be tuned finely from 350 Kelvin to the lowest energy of collision of almost 10 milli-Kelvin. With this capability, they have noticed orbiting resonance characteristics in the rate of reaction. More significantly, this technique enables the examination of cold collision of the molecular/atomic beams with the energy of collision spanning up to six orders of magnitude, as displayed in Figure 3.15. It, therefore, bridges a huge gap of energy between traditional studies of the crossed-molecular beam collisions and the studies of ultracold collision, which usually have been out of reach for several decades in the studies of collision.

With this method, the effect of isotope on the rate of reaction has been studied (Lavert et al., 2014). Partial waves of the collision with diverse isotopologs trigger scattering resonances at diverse energies of collision, leading to different rates of reaction for diverse isotopes in collision range of temperature from 10 mK-100 K. For a similar isotope of the hydrogen (H) molecule reacting with He, the constant of reaction rate with He(2 $^3P_2$) is generally more than 2 order of magnitude as compared to reaction with He(2 $^3P_2$). It exhibits significant dependency on the state of rotation as is displayed in Figure 3. 16 (Landau & Moiseyev, 2016). The constant reaction rate of paraH_2 molecules with Helium (2 $^3P_2$) trails theoretical universal Langevin estimation with the $E^{1/6}$ scaling down to 0.8 Kelvin, below which the constant of rate falls quite below $E^{1/6}$ scaling and ultimately slow by 50% at 10 milli-Kelvin lowest energy of a collision.

3.4.3. Molecular Synchrotron

This can normally be viewed as the discrepancy of crossed beam apparatus with 0 degrees or 180-degree angle and is analogous to the fashion of the slow molecular beam striking with particles in the trap.

One dissimilarity is that the molecular packets can be loaded continuously, generating a large number density of the molecules (Van & Meijer, 2009). As a field for low-energy molecular collision research, molecular synchrotron provides multiple desirable benefits. Firstly, the collision partners can travel in a similar direction, occasioning the small relative velocity and therefore the low collision energy. If beams that travel with a similar speed as the molecules stored are chosen, in principle, 0 energy of collision can be obtained (Raghavachari et al., 1989; Ghigo et al., 2004).

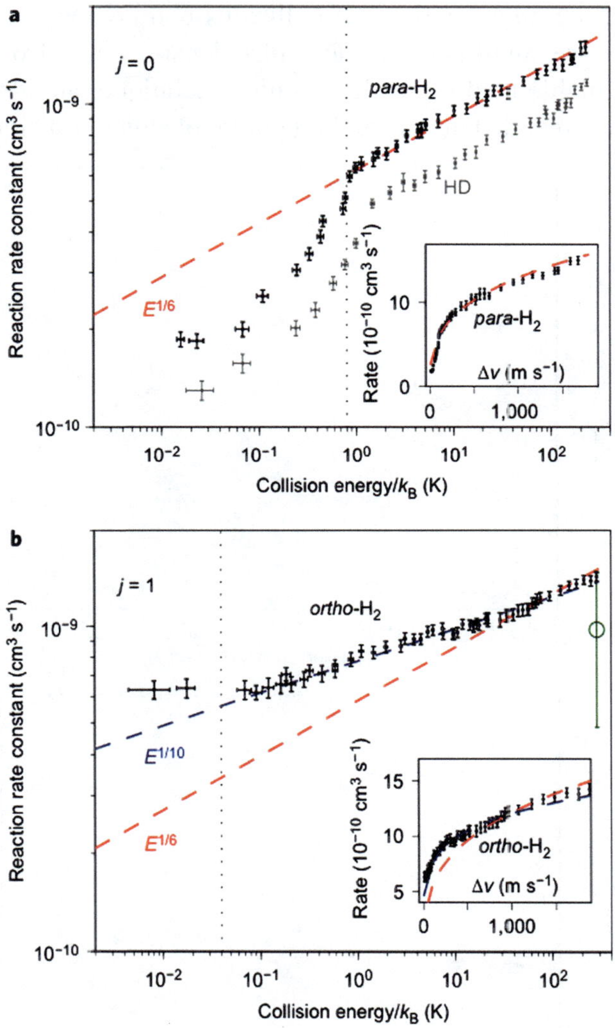

Figure 3.16. Measurement of reaction rate for the hydrogen molecules in (i) ground and (ii) excited rotational states with Helium (2 3P2) from 10 mK-300 K.

Source: https://www.nature.com/articles/nchem.2359.

Secondly, as the cross-sections of collisions comprising neutral atoms or molecules or are small and the molecular densities are also low from the presently available cooling or deceleration technique, an adequately large sensitivity is demanded understanding cold molecular collisions.

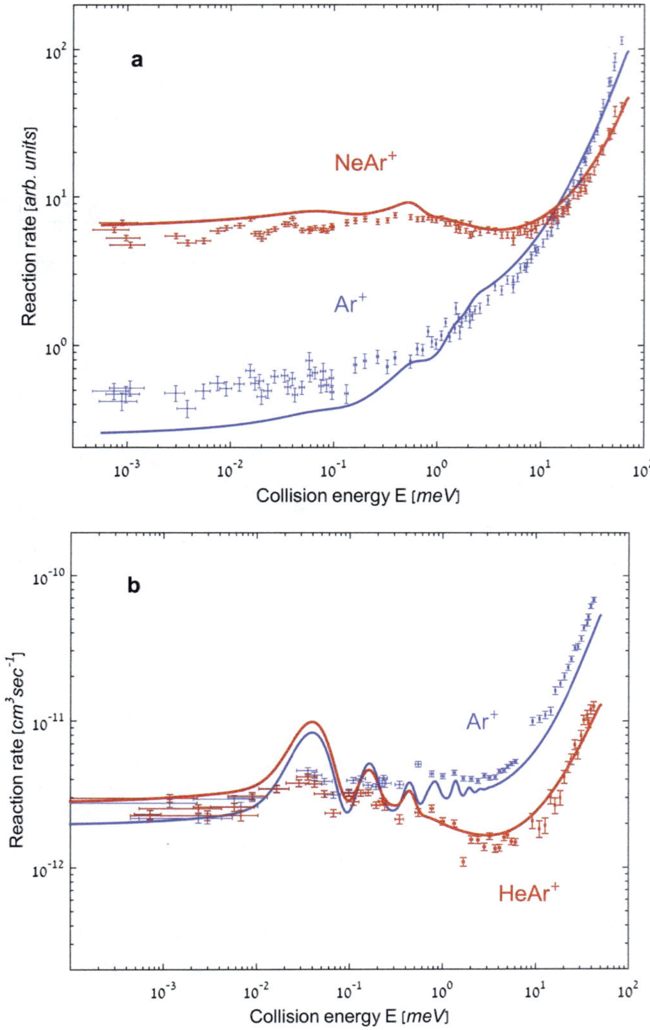

Figure 3.17. Measurement of reaction rate of Penning ionization (blue) and associative ionization (red) in reaction He(3S) + Ar (b) and Ne(3P) + Ar (a). The solid-lines represent numerical simulations.

Source: https://www.tandfonline.com/doi/abs/10.1080/00268976.2019.1594421.

As molecular synchrotron can usually store molecular packs for several round-trips, they can also interact recurrently with the collision partners at distinct positions and at well-defined times. The longer the packs are kept before finding, the more of the molecules are lost because of (in)elastic collisions gather over the events of collision, therefore collision signals are gathered, and the reactivity to collisions will be enhanced strongly. Considering there exist *n* packets rotating clockwise with a similar speed in a synchrotron comprising of *N* segments, every packet of the molecules collide with co-propagating injection packet once after every round trip. After *m* number of round trips, it could have intermingled *mn* times. With a surplus factor $\sqrt{n}$ because of the summed intensity of *n* packets, the complete gain in sensitivity will be $mn\sqrt{n}$, which might amount to the factor of 100 to 1000 (Poel & Bethlem, 2018).

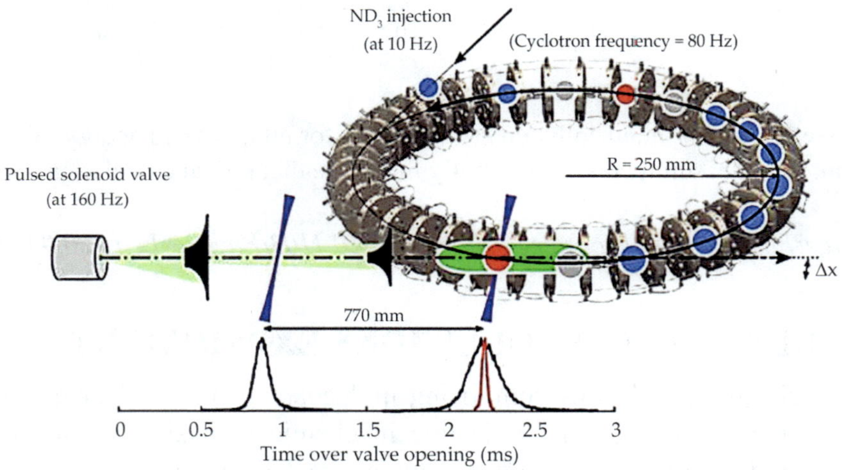

Figure 3.18. A graphic view of the beamline and synchrotron. Supersonic packs of Argon (green) from the chilled Even-Lavie valve meet co-propagating probe ND_3 packs (red) kept in synchrotron on each round-trap, whereas reference ND_3 packets (blue) provide the instantaneous measurement of background loss.

Source: https://journals.aps.org/prl/abstract/10.1103/PhysRevLett.120.033402.

Molecular synchrotron was established by Heiner et al. (2013), which disrupts the molecular storage ring (Crompvoets et al., 2001) into two half-circles and then switches the voltages on hexapole electrodes in a way that molecules are clustered together as the molecules fly through the opening between the two half rings. In this manner, the molecular packs are both longitudinally and transversely confined, therefore averting spreading out beside the direction of flight and maintaining adequate density in order to carry out collision experimentations. In this sample synchrotron, the molecules were preserved together for around a hundred round trips, and infusion of various molecular packs was demonstrated.

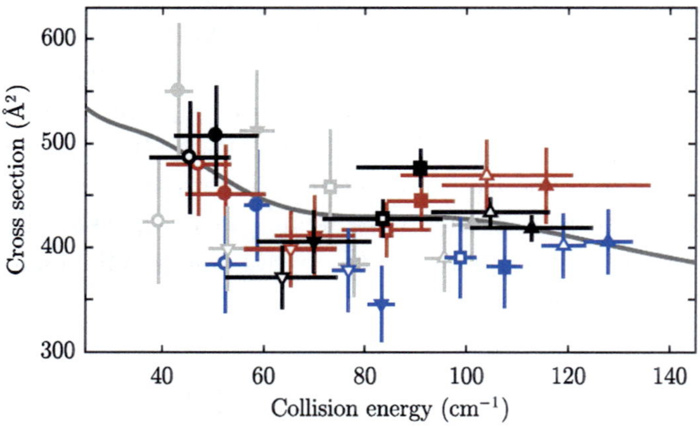

Figure 3.19. Combined collision cross-section for ND_3 + Ar vs energy of a collision. The measurements are normally fit for the theoretical calculation.

Source: https://journals.aps.org/prl/abstract/10.1103/PhysRevLett.120.033402.

3.5. ULTRA-COLD COLLISIONS EXPERIMENT

Comprehending collisions comprising molecules at the level of quantum has been the long-standing objective in chemical physics. As inelastic, elastic, and reactive collision cross-sections are delicate to the particulars of both long-range and short-range part of molecular interaction potential, the ultra-cold molecular collision provides a perfect probe of PES, conversely, the scattering resonances measurement provides a precise probe of PESs. In this viewpoint, the ultra-cold molecules provide great chances to learn molecular strikings in the quantum region, since the de Broglie wavelength of collision partners is larger as compared to a range of the molecular interaction perspective at ultra-cold temperature and therefore only the least partial wave of the comparative orbital angular momentum governs the process of collision. As the first relationship of ultra-cold atoms, collisions between the atoms and the molecules have been studied widely (Thorsheim et al., 1987).

3.5.1. Vibrationally Excited Molecule

3.5.1.1. Homonuclear Diatomic Molecule

Dimers of the fermions have usually been perceived with lifetimes longer as compared to the scales of times for thermalization and elastic collisions, showing extraordinary stability against the molecule-molecule and atom-molecule collisions. The $^{40}K_2$ molecules are produced by scanning the almost equal mixture of |9/2,−9/2i, and |*f* = 9/2, m_f = −5/2i spin state over the Feshbach resonance at around 224.21 G adiabatically at the ramping rate of nearly 40μsG^{-1} [244]. The lifetime of molecules measured has been only one ms owing to the vibrational quenching, though it increased intensely as Feshbach resonance is reached, which is validated in the subsequent experiment [247]. The molecules are made from degenerate atoms of ^{40}K in an equal mixture of |9/2,−9/2i, and |9/2,−7/2i state, and the lifetime upsurges from ~ one millisecond to greater than 100 milliseconds.

The molecules of Li_2 show the analogous collisional stability close to the Feshbach resonance. Kevin E. Strecker et al. developed molecules of Li_2 in the lowest-bound |*v* = 38i level of vibration of the $X^1\Sigma^+_g$ singlet-state from the degenerate two components Fermi gas |1/2,1/2i and |1/2,−1/2i by adiabatic passage via the slim Feshbach resonance at nearly 543 G. The molecules are generated with a productivity of 50% and remain restricted in the optical trap for nearly one second. Likewise, J. Cubizolles et al. develop weakly bound molecules of Li_2 by sweeping the magnetic field across the wide Feshbach resonance at nearly 810 G with 85% efficiency (Cubizolles et a., 2003). The deterioration of Li_2 molecules exhibits the lifetime ≃ 500*ms* at almost 689 G, whereas only 20 milliseconds at 636 G, displaying strikingly diverse collisional relaxation. Jochim et al. made a pure specimen of feebly bound molecules of Li_2 from 3-body recombination in the 50–50 spin blend in the least two spin states at the field of almost 690 G, and perceive the strong coupling amongst the molecules and the atomic gas. The long lifespan of the molecules signifies the large elastic rate of collision amongst the particles. The collisional stability is explicated by fermionic suppression of the vibrational quenching in the collisions of the molecule and triggers the immediate creation of the molecular Bose-Einstein condensation (BEC) after evaporative cooling. (Greiner et al., 2003). Tout T. Wang et al. have detected diverse collisional behavior in collisions of the Li_2 molecules in a highest vibrational state having free Lithium atoms, compared to Li_2 + Na and Li_2 + Li_2 collisions. The 2-body inelastic coefficients of decay for the latter 2 collisions match with calculations from the model of quantum Langevin for general collisions of the ultracold molecules, which supposes the unit possibility of loss at short-range and the rate of decay reliant on the van der Waals interaction (long-range) amongst collisions partners.

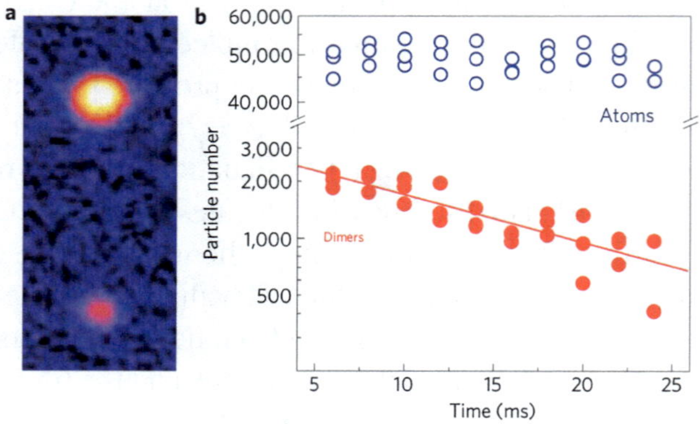

Figure 3.20. Measuring the rate of loss of Cs2 dimers. (i) Absorption image of an atom-dimer blend after discharged from the snare and the Stern-Gerlach separation. (b) The time evolution of the number of dimers and atoms at 35 G. The dimers loss is fitted with the exponential decay curve.

Source: https://www.nature.com/articles/nphys1203.

Though the 2-body coefficient of decay for the previous collision is very small, more than ten times smaller as compared to the general predictions, which is described by a low density of the accessible decay states in systems of the light atoms (Balakrishnan et al., 2011).

On the contrary, dimers of the bosons exhibit a rapid decay through inelastic collision, so that the quantum degenerated molecular clouds can survive for a few ms. It was stated the formation of state-chosen molecules of Rb_2 in the rubidium BEC having coherent free-bound encouraged Raman transitions whose shapes of line imitate the collisional behavior of atoms and molecules. The inelastic collisions of atom-molecule outcome in the finite lifetime of molecules, whereas the elastic interactions amongst the atomic and molecule condensate give upsurge to the mean-field influence to the chemical potential of the molecule or atom, which triggers the shift of resonance Raman transition. These interactions are accountable for the widening of Raman transition shapes of line.

3.5.1.2. Heteronuclear Diatomic Molecule

Researchers perceived strong inelastic collisions of the ultra-cold Rubidium Caesum with background Rb or Cr atoms, which trigger dramatic lessening of the lifetime of molecules. The vibrationally excited Rubidium Caesum molecule in its $a^3\Sigma^+$ ground electronic state is created through photo association of the laser-cooled ^{133}Cs and ^{85}Rb atoms. The inelastic constant of scattering rate displays a lack of dependency on the molecular energy of binding, signifying the particulars of short-range interaction prospective are insignificant, consistent with the model of Langevin. J. J. Zirbel et al. made ultra-cold feebly bound fermionic molecules of 40 K 87 Rb using the Feshbach resonance between the atoms of K |9.2,−9/2i and Rb |1,1i at $B_0 = 546.7G$ and examine their stability in collisions with fermionic ^{40}K, bosonic ^{87}Rb, or distinct ^{40}K |9/2,−7/2i atoms (Balakrishnan et al., 2011).

However, improved rates of decay for a greater than $1000a_0$ are witnessed when colliding with the atoms of bosonic ^{87}Rb. Researchers examined collisions of the ultra-cold polar molecules of LiCs and ultra-cold atoms of cesium in the optical dipole trap. The molecules of LiCs are generated by photo association of the atoms of lithium and cesium the $B^1\Pi$ excited-state trailed by impulsive emission to the lowest triplet-sate $a^3\Sigma^+$and the $X^1\Sigma^+$ ground state.

3.5.1.3. Collision Amongst Feshbach Molecules

There are studies on the collisions of ultra-cold molecule-molecule perceived magnetically adjusted collision resonances for the ultracold molecules of Cs_2 stored in the CO_2-laser trap. The perceived double-peak structure strongly recommends that the presence of bound states of 2 molecules of Cs_2 or tetramer states of Cs_4 in collisions amongst ultra-cold Cs_2 molecules, tempting Feshbach-like couplings to the inelastic decay channels. Researchers examined the Cs_2 halo dimers loss because of inelastic dimer-dimer collision in the pure trapped specimen of Feshbach molecules of Cs_2. The dependencies of coefficient relaxation rate on scattering temperature and length are measured. Scholars currently stated the opinion of the dimer-dimer inelastic collisions resonance for the ultra-cold Feshbach molecules of $^{23}Na^{87}Rb$. The resonance demonstrates itself as the noticeable inelastic peak loss of the dimers when the interspecies length of scattering is adjusted and is accredited to the overpassing of the dimer-dimer threshold with the tetramer state. The coefficients of the rate of loss also show the temperature dependency.

3.5.2. Ground State Molecule

3.5.2.1. Field-Free

The 1st ultracold collision between rovibrational ground-state molecules and atoms is studied recently, where the nature of the comparative motion of reactants is imitated in the quantum verge behavior, and the statistics become unequivocally significant.

When the molecules are made in the equal blend of two hyper-fine states, the reaction continues rapidly and the temperature-independent rate of loss is witnessed, where the molecules' loss is found because of the hindrance-free chemical reaction with the atoms of potassium.

To define the experimental outcomes for the reaction of the ultra-cold molecules of KRb, S. Ospelkaus et al. have utilized both MQDT and QT models to examine the rate of loss of the rovibronic ground state molecules of 40 87 K Rb molecules. The contrast of the calculation outcomes based on every model to the measurements done experimentally is demonstrated in Figure 3.21. The data of experiments is in agreement with the threshold laws of Wigner, where the rate of reaction slows down at the low temperatures as the rate of tunneling is proportional to temperature. In the current experiment, the 2-body loss coefficients of the molecules of KRb follow the anticipated MQDT trend strictly with temperature $T/T_F > 0.6$, whereas exhibiting strong deviations with temperature $T/T_F \leq 0.6$ at all of the temperatures (Marco et al., 2019).

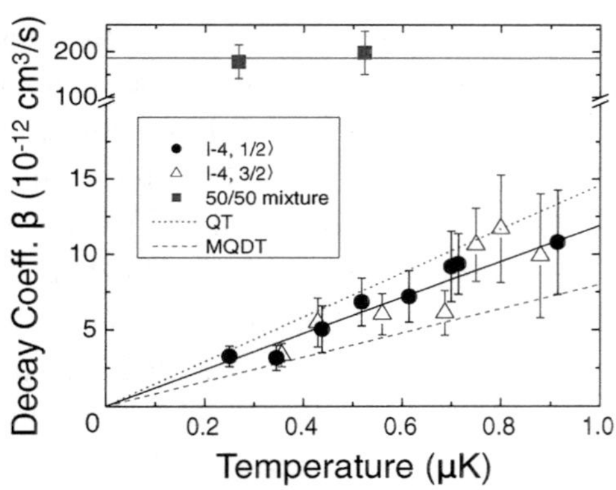

Figure 3.21. Coefficient of loss rate of the $^{40}K^{87}Rb$ vs temperature.

Source: https://science.sciencemag.org/content/327/5967/853.

Note: Experimental data for the single constituent gas of molecules in the spin states | m_I^{Rb} = 1/2i, m^K_I = –4 (solid circles), | m_I^{Rb} = 3/2i, m^K_I = –4 (open triangles), and for the 50/50 blend of these 2 spin states (solid squares), are matched with extrapolation of MQDT (dashed line) and QT (dotted line) model. The linear fit (solid line) to | – 4, 1/2i produces the temperature-dependent rate of loss to be nearly 1.2(±0.3) × 10^{-5} cm^3s^{-1}/K.

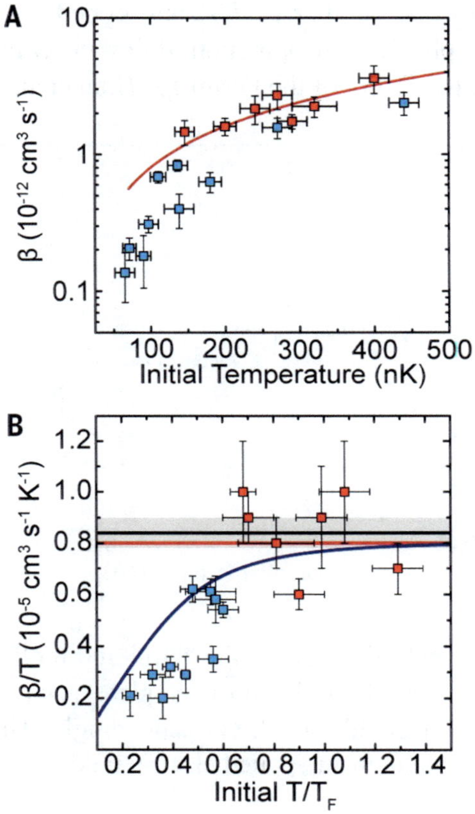

Figure 3.22. Temperature dependency of the reaction rate constants. (i) The rate of reaction constant β at several initial temperatures. (ii) Temperature normalized rate of reaction constants as the function of degeneracy. The solid blue points relate to T/TF ≤ 0.6 and the solid red points relate to T/TF > 0.6. Solid blue, red, and black lines signify average β/T for the T/TF > 0.6, MQDT computations, and the average relative density fluctuations, correspondingly.

Source: https://science.sciencemag.org/content/363/6429/853.

Furthermore, temperature normalized rate of reaction constants β/T is persistent above temperature T/T_F = 0.6, in outstanding agreement with forecasted MQDT value of the β/T = 0.8×10^{-5} cm^3s^{-1}/K, while β/T displays the strong deviation from the threshold law of Wigner and the MQDT model trends at temperature $T/T_F \leq 0.6$, as demonstrated in Figure 3.22. The perceived reduction of β/T, demonstrating the suppressed rate of collision beyond the estimation by the threshold law of Wigner, is accredited to the decreased density fluctuation, i.e., decreased probability of finding two molecules within the short distance of one another as T/T_F is dropped due to anti-bunching, which is the similar physical phenomenon giving upswing to Pauli pressure and decreased compressibility of the Fermi gas (Lau et al., 2016).

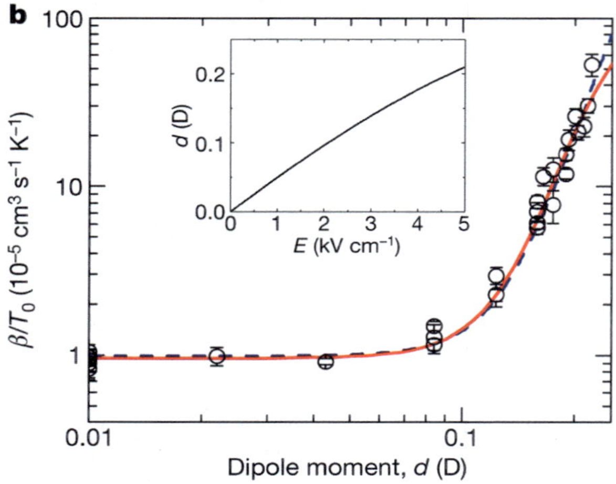

Figure 3.23. 2-body inelastic loss for the fermionic molecules of KRb. The solid line and dashed line exhibit the fit to complete quantum scattering model and simple quantum threshold model, correspondingly. The inset exhibit the calculated dependency of d on an applied electric field.

Source: https://www.nature.com/articles/nature08953.

To test the sultry collision hypothesis (Makrides et al., 2015; Lau et al., 2016), which says that pairs of the molecules develop long-lived collision networks and is thought to be the main mechanism for the ground state molecules loss from the snare, Recently it investigated experimentally collisional losses of the optically stuck sample of the ground state molecules of $^{87}Rb^{133}Cs$. The molecules are made in the single hyperfine level of $X^1\Sigma^+$ rovibrational ground state through stirred Raman adiabatic passage. The losses of collisions of the non-reactive molecules are measured. The molecule loss is discovered consistent with the anticipation for intricate arbitrated collisions, but the rate of loss is lower as compared to the restriction of universal loss.

3.5.2.2. External Field

The tunability of the chemical reactivity of the polar molecules has been considered under the external field. It was perceived the onset of hyper-fine varying strikings when the magnetic field is inclined so that the molecules of $^{87}Rb^{133}Cs$ are not in the hyper-fine ground state | *J*" = 0, *v*" = 0, M_{tot} = 5i, and establish that the 2-body rate of loss coefficient decrease by the order of magnitude when the magnetic field is augmented to more than ninety G. It was validated that uncertain exterior electric fields can radically change the dipole interactions amongst the fermionic molecules of KRb. In an ultra-cold dipolar collision, the loss of fermionic molecules of KRb molecules because of ultra-cold chemical reactions KRb + KRb → K_2 + Rb_2, and the power-law dependency of the rate of loss on the tempted electric dipole moment has been perceived, as displayed in Figure 3.23 (Makrides et al., 2015). At the low electric fields where $d < 0.1D$, no important variation to the rate of loss has been noticed. For the higher electric fields, the rate of loss β/T_0 quickly increases well over the order of magnitude by the relation $d = 0.2D$, and the coefficient of rate can be fixed to $\beta/T_0 \propto d^p$ and the value of $p = 6.1 \pm 0.8$. The normalized fractional rate of heating $T'/T_0^2 n$ trails the same style with the rate of loss, agreeing with the simple thermodynamic model supposing that the heating is caused directly by inelastic loss. The dipolar interaction exhibit the nature of spatial anisotropy from parametric driving computations under the external electric fields (Sowokinos, 2001). The case in which primarily $T_z > T_x$, T_z, and T_x reach one another and the scale of time for the re-thermalization reduces steeply as d upsurges. whereas T_z and T_x don't equilibrate when a gas primarily has $T_z < T_x$. This recommends that it is essential to shield the molecules from heating and strong inelastic loss and restricting them in the array of pancake-shaped traps in the 1-D optical lattice arrangement would be an encouraging route to realize the quantum gas of the polar molecules.

3.6. APPLICATION OF THE ULTRA-COLD MOLECULES

Presently the research of ultra-cold molecules has appeared as the fast-growing field because of their probable application in various subjects varying from an accurate measurement of the essential physical constants to quantum simulation and calculation, to the vital experiment of Charge-Parity-Time symmetry and the research of physics above the Standard model and ultra-cold chemistry (Carr et al., 2009; Banerjee, 2012). Nevertheless, cooling the molecules directly to an ultracold region from room temperature or above remains the major challenge due to the intricate molecular internal degrees of freedom (DOF). So far, the ultra-cold molecules can generally be made by a relationship of laser cooling ultracold atoms and electro-optic Sisyphus cooling (Shuman et al., 2010; Zeppenfeld et al., 2012).

3.6.1. Evaporative Cooling

Evaporative cooling of the thermal distribution is decreasing the collective temperature by selectively eliminating particles having energies much greater as compared to the average total energy for each particle. In the existence of elastic collisions, high-energy extension is recurrently repopulated and be cropped, realizing elimination of energy at price in the particle number. Evaporative cooling has accomplished great success in the understanding of quantum degenerated gases, and applicability to the molecules has been discussed theoretically. Ultracold and cold collisions of the molecules of NH have been examined without or with an external magnetic field, representing a large ratio of inelastic to elastic cross-section and predictions for evaporative cooling (Wallis et al., 2011). Conversely, it is discovered that ultra-cold reaction of the magnetically stuck $NH(X^3\Sigma^-$ radicals are compelled by the short-range collisional mechanism, and the magnitude of the reactive cross-section is feebly dependent on the strength of the magnetic field, which specify chemical reactions might cause more snare loss as compared to inelastic spin-varying collisions, making the process of evaporative cooling more difficult. Former experimental work on ND_3 and KRb molecules recommend that evaporative cooling is not favorable for these kinds of molecules. For the OH molecules, in spite of an open shell arrangement and the large anisotropy in OH-OH interaction potential, the theoretical work found strong suppression of the inelastic rate of collision constant amongst OH molecules in the magnetic field. Benjamin K. Stuhl et al. realized experimentally the process of evaporative cooling of OH molecules in their ground-state The structure of Λ-doublet utilizing the radio frequency (RF) knife to consecutively lower the depth of trap (Deiglmayr et al., 2008; Lang et al., 2008).

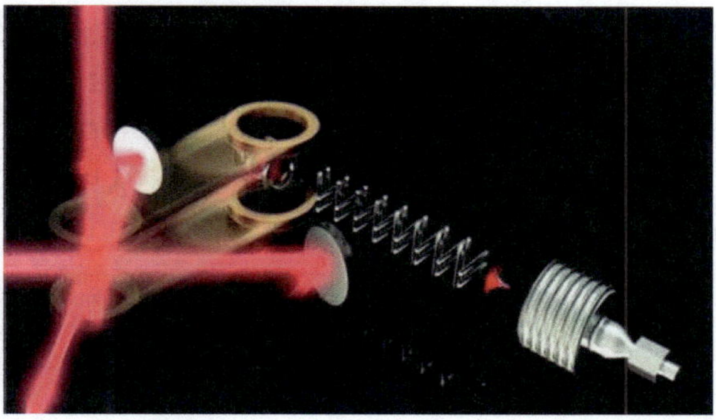

Figure 3.24. The atoms of Rb are cooled and stuck at the intersection of the laser beams, whereas the supersonic beam of the ND_3 molecules is made by the pulsed valve, trapped, and slowed utilizing inhomogeneous time-changing electric fields generated by an array of the high voltage electrodes. The duo of antiHelmholtz coils comprising atomic trap is interpreted along with the linear translation phase such that the atomic trap intersects with an electrostatic molecular trap.

Source: https://journals.aps.org/prl/abstract/10.1103/PhysRevLett.106.193201.

3.6.2. Sympathetic Cooling

The process is thermalization of the candidate species in the heat bath of ultra-cold particles. It has accomplished great achievement in cooling atoms and ions. Because of this reason, its use to molecules was expected naturally and therefore many efforts have been dedicated for several different atom-molecule systems by the theorists in the last two decades, like NH_3–Rb, OHRb, H_2-Rare gas, NH-Alkali atom, OH-H, and NH-H, N-NH, LiH-Li, YbF-He, CaH-Mg, and CaH-Li, CaF-Rb, and CaF-Li, SrOH-Li, and SrF-Rb (Morita et al., 2019).

The usual method of sympathetic cooling of the molecules utilizing ultracold atoms would be the intersecting of the magneto-optical snare of atoms with the molecular trap supposing encouraging rate constants ration for spin relaxation and elastic scattering between molecules and atoms. Conversely, the large optical entrance of this magneto-optical snare and enclosure arrangement of electrostatic or the magneto-static snare for molecules, along with the small volume of the molecular trap, makes a substantial technical task for an experimental realization. Up to now, the inelastic collisions between Rb and ND_3 have been examined experimentally. Ultracold atoms of Rb are made in a distinct chamber primarily because of the necessity for large access. The atoms of rubidium are laser-cooled and restricted in the magneto-optical snare, then these atoms are loaded into the magnetic trap and moved to the trapping regime of ND_3 by transferring magnetic coils along with the linear translation phase, as is displayed in Figure 3.24.

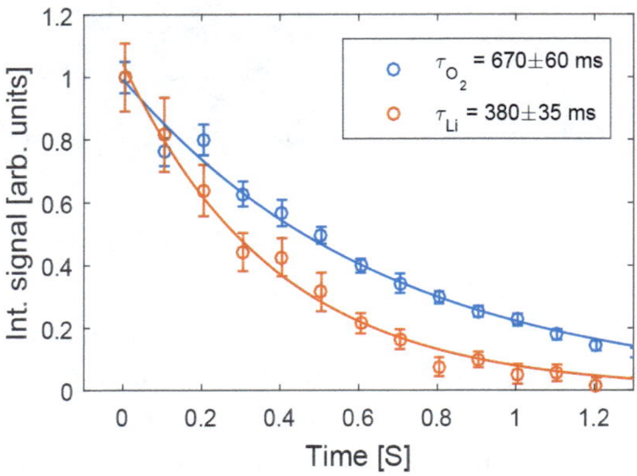

Figure 3.25. The measured lifetime of the stuck O_2 molecules in a magnetic trap for diverse trap depths.

Source: https://www.nature.com/articles/s41586–019–1446–2.

Currently, the Narevicius group used a different method of moving the snare decelerator - to intersect molecules with the alkali atoms (Akerman et al., 2015). A similar method with a different configuration of the coil has been validated. The principle has formerly been made practical to decelerate the polar molecules with moving electric potential both in free space and on a chip.

Very currently, a group studies the properties of collision for the molecule of oxygen alone and also for the molecules of oxygen with Li atom inside the superconducting magnetic snare by measuring the lifetimes, as displayed in Figure 3.25. The molecules of O_2 alone can be stuck for more than 90 seconds with a lifetime of $\tau \approx 52$ seconds when these molecules are loaded into the shallow snare of 50 mK, constant with an expected background rate of collision, demonstrating the deficiency of intermolecular collisions. With a growing depth of trap, clear deviance from the exponential decay has been detected and the O_2–O_2 2-body collisions are pronounced more. Nearly 9 seconds 2-body lifetime is expected at 800 mK depth of trap. In Figure 3.26, stuck Li atoms also show the exponential decay with the lifetime of $\tau \approx 14$ seconds when loading into a trap without the molecules of oxygen.

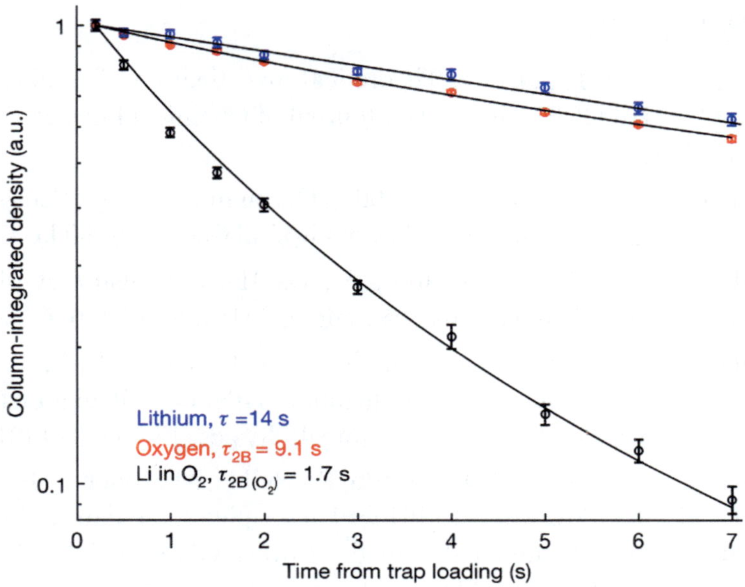

Figure 3.26. A measured lifetime of the stuck Li atoms and O_2 molecules in a magnetic trap.

Source: https://www.nature.com/articles/s41586–019–1446–2.

When Li is co-loaded with the O_2 the lithium signal displays the rapid non-exponential decay with a 1.7 second lifetime. The lifetime of O_2 doesn't exhibit any change because of the addition of Li, approving stuck density of lithium is much lower as compared to that of oxygen (Ospelkaus et al., 2008).

REFERENCES

1. Alexander, M. H., (1998). Theoretical investigation of weakly-bound complexes of O(^{3}P) with $_{H2}$. The Journal of Chemical Physics, 108(11), 4467–4477.
2. Althorpe, S. C., & Clary, D. C., (2003). Quantum scattering calculations on chemical reactions. Annual Review of Physical Chemistry, 54(1), 493–529.
3. Balakrishnan, N., & Dalgarno, A., (2001). Chemistry at ultracold temperatures. Chemical Physics Letters, 341(5, 6), 652–656.
4. Balakrishnan, N., Quéméner, G., Forrey, R. C., Hinde, R. J., & Stancil, P. C., (2011). Full-dimensional quantum dynamics calculations of H_2–H_2 collisions. The Journal of Chemical Physics, 134(1), 014301.
5. Banerjee, J., Rahmlow, D., Carollo, R., Bellos, M., Eyler, E. E., Gould, P. L., & Stwalley, W. C., (2012). Direct photoassociative formation of ultracold KRb molecules in the lowest vibrational levels of the electronic ground state. Physical Review A, 86(5), 053428.
6. Baranov, M. A., (2008). Theoretical progress in many-body physics with ultracold dipolar gases. Physics Reports, 464(3), 71–111.
7. Barnwell, J. D., Loeser, J. G., & Herschbach, D. R., (1983). Angular correlations in chemical reactions. Statistical theory for four-vector correlations. The Journal of Physical Chemistry, 87(15), 2781–2786.
8. Bergeat, A., Onvlee, J., Naulin, C., Van, D. A. A., & Costes, M., (2015). Quantum dynamical resonances in low-energy CO (j= 0)+ He inelastic collisions. Nature Chemistry, 7(4), 349–353.
9. Bernstein, R. B., (1979). Introduction to atom-molecule collisions: The interdependency of theory and experiment. In: Atom-Molecule Collision Theory (pp. 1–43). Springer, Boston, MA.
10. Berteloite, C., Lara, M., Bergeat, A., Le Picard, S. D., Dayou, F., Hickson, K. M., & Costes, M., (2010). Kinetics and dynamics of the S (D 2 1) + H_2→ SH + H reaction at very low temperatures and collision energies. Physical Review Letters, 105(20), 203201.
11. Brandao, J., Mogo, C., & Silva, B. C., (2004). Potential energy surface for $H_2O(^3A'')$ from accurate ab initio data with inclusion of long-range interactions. The Journal of Chemical Physics, 121(18), 8861–8868.
12. Buchachenko, A. A., Chałasiński, G., & Szczęśniak, M. M., (2007). Interactions of lanthanide atoms: Comparative ab initio study of YbHe, Yb_2 and TmHe, TmYb potentials. The European Physical Journal D, 45(1), 147–153.

13. Campbell, W. C., Tscherbul, T. V., Lu, H. I., Tsikata, E., Krems, R. V., & Doyle, J. M., (2009). Mechanism of collisional spin relaxation in Σ 3 molecules. Physical Review Letters, 102(1), 013003.

14. Campbell, W. C., Tsikata, E., Lu, H. I., Van, B. L. D., & Doyle, J. M., (2007). Magnetic trapping and Zeeman relaxation of NH (X Σ− 3). Physical Review Letters, 98(21), 213001.

15. Carr, L. D., DeMille, D., Krems, R. V., & Ye, J., (2009). Cold and ultracold molecules: science, technology and applications. New Journal of Physics, 11(5), 055049.

16. Chalasinski, G., & Szczesniak, M. M., (1994). Origins of structure and energetics of Van Der Waals clusters from ab initio calculations. Chemical Reviews, 94(7), 1723–1765.

17. Chefdeville, S., Kalugina, Y., Van De, M. S. Y., Naulin, C., Lique, F., & Costes, M., (2013). Observation of partial wave resonances in low-energy O_2–H_2 inelastic collisions. Science, 341(6150), 1094–1096.

18. Chefdeville, S., Stoecklin, T., Bergeat, A., Hickson, K. M., Naulin, C., & Costes, M., (2012). Appearance of low energy resonances in CO–para-H_2 inelastic collisions. Physical Review Letters, 109(2), 023201.

19. Chin, C., Kraemer, T., Mark, M., Herbig, J., Waldburger, P., Nägerl, H. C., & Grimm, R., (2005). Observation of Feshbach-like resonances in collisions between ultracold molecules. Physical Review Letters, 94(12), 123201.

20. Crompvoets, F. M., Bethlem, H. L., Jongma, R. T., & Meijer, G., (2001). A prototype storage ring for neutral molecules. Nature, 411(6834), 174–176.

21. Cubizolles, J., Bourdel, T., Kokkelmans, S. J. J. M. F., Shlyapnikov, G. V., & Salomon, C., (2003). Production of long-lived ultracold Li_2 molecules from a fermi gas. Physical Review Letters, 91(24), 240401.

22. De Marco, L., Valtolina, G., Matsuda, K., Tobias, W. G., Covey, J. P., & Ye, J., (2019). A degenerate fermi gas of polar molecules. Science, 363(6429), 853–856.

23. Deiglmayr, J., Grochola, A., Repp, M., Mörtlbauer, K., Glück, C., Lange, J., & Weidemüller, M., (2008). Formation of ultracold polar molecules in the rovibrational ground state. Physical Review Letters, 101(13), 133004.

24. Deiglmayr, J., Repp, M., Wester, R., Dulieu, O., & Weidemüller, M., (2011). Inelastic collisions of ultracold polar LiCs molecules with caesium atoms in an optical dipole trap. Physical Chemistry Chemical Physics, 13(42), 19101–19105.

25. DeMille, D., (2002). Quantum computation with trapped polar molecules. Physical Review Letters, 88(6), 067901.
26. Dhont, G. S., Van, L. J. H., Groenenboom, G. C., & Van, D. A. A., (2005). Ab initio calculation of the NH (Σ− 3)− NH (Σ− 3) interaction potentials in the quintet, triplet, and singlet states. The Journal of Chemical Physics, 123(18), 184302.
27. Drayna, G. K., Hallas, C., Wang, K., Domingos, S. R., Eibenberger, S., Doyle, J. M., & Patterson, D., (2016). Direct Time-domain observation of conformational relaxation in gas-phase cold collisions. Angewandte Chemie International Edition, 55(16), 4957–4961.
28. Dubois, B., Hampel, H., Feldman, H. H., Scheltens, P., Aisen, P., Andrieu, S., & Washington, D. C., (2016). Preclinical Alzheimer's disease: Definition, natural history, and diagnostic criteria. Alzheimer's & Dementia, 12(3), 292–323.
29. Dunning, Jr. T. H., (1989). Gaussian basis sets for use in correlated molecular calculations. I. The atoms boron through neon and hydrogen. The Journal of Chemical Physics, 90(2), 1007–1023.
30. Dupeyrat, G., Marquette, J. B., & Rowe, B. R., (1985). Design and testing of axisymmetric nozzles for ion-molecule reaction studies between 20° K and 160° K. The Physics of Fluids, 28(5), 1273–1279.
31. Eisert, J., Friesdorf, M., & Gogolin, C., (2015). Quantum many-body systems out of equilibrium. Nature Physics, 11(2), 124–130.
32. Frishman, A., Hoffman, D. K., & Kouri, D. J., (1997). Distributed approximating functional fit of the H_3 ab initio potential-energy data of Liu and Siegbahn. The Journal of Chemical Physics, 107(3), 804–811.
33. Garcia, E., & Lagana', A., (1985). A new bond-order functional form for triatomic molecules: A fit of the befh potential energy. Molecular Physics, 56(3), 629–639.
34. Ghigo, G., Roos, B. O., & Malmqvist, P. Å., (2004). A modified definition of the zeroth-order Hamiltonian in multiconfigurational perturbation theory (CASPT2). Chemical Physics Letters, 396(1–3), 142–149.
35. Greiner, M., Regal, C. A., & Jin, D. S., (2003). Emergence of a molecular Bose-Einstein condensate from a fermi gas. Nature, 426(6966), 537–540.
36. Hapka, M., Żuchowski, P. S., Szczęśniak, M. M., & Chałasiński, G., (2012). Symmetry-adapted perturbation theory based on unrestricted

Kohn-sham orbitals for high-spin open-shell Van Der Waals complexes. The Journal of Chemical Physics, 137(16), 164104.

37. Harich, S. A., Dai, D., Wang, C. C., Yang, X., Der Chao, S., & Skodje, R. T., (2002). Forward scattering due to slow-down of the intermediate in the H+ HD→ D+ H_2 reaction. Nature, 419(6904), 281–284.
38. Henson, A. B., Gersten, S., Shagam, Y., Narevicius, J., & Narevicius, E., (2012). Observation of resonances in penning ionization reactions at sub-kelvin temperatures in merged beams. Science, 338(6104), 234–238.
39. Herbst, E., & Yates, Jr. J. T., (2013). Introduction: Astrochemistry. Chemical Reviews, 113(12), 8707–8709.
40. Herbst, E., (1995). Chemistry in the interstellar medium. Annual Review of Physical Chemistry, 46(1), 27–54.
41. Herschbach, D. R., (1987). Molecular dynamics of elementary chemical reactions (Nobel lecture). Angewandte Chemie International Edition in English, 26(12), 1221–1243.
42. Herschbach, D., (2009). Molecular collisions, from warm to ultracold. Faraday Discussions, 142, 9–23.
43. Ho, T. S., & Rabitz, H., (2000). Proper construction of ab initio global potential surfaces with accurate long-range interactions. The Journal of Chemical Physics, 113(10), 3960–3968.
44. Hollebeek, T., Ho, T. S., & Rabitz, H., (1999). Constructing multidimensional molecular potential energy surfaces from ab initio data. Annual Review of Physical Chemistry, 50(1), 537–570.
45. Hudson, E. R., Gilfoy, N. B., Kotochigova, S., Sage, J. M., & DeMille, D., (2008). Inelastic collisions of ultracold heteronuclear molecules in an optical trap. Physical Review Letters, 100(20), 203201.
46. Hudson, E. R., Lewandowski, H. J., Sawyer, B. C., & Ye, J., (2006). Cold molecule spectroscopy for constraining the evolution of the fine structure constant. Physical Review Letters, 96(14), 143004.
47. Hummon, M. T., & Tscherbul, T. V., (2011). In: Los, J. K., Lu, H. I., Tsikata, E., Campbell, W. C., Dalgarno, A., & Doyle, J. M., (eds.), Phys. Rev. Lett. (Vol. 106, p. 053201.
48. James, P. L., Sims, I. R., Smith, I. W., Alexander, M. H., & Yang, M., (1998). A combined experimental and theoretical study of rotational energy transfer in collisions between NO (X 2 Π 1/2, v= 3, J) and He, Ar and N_2 at temperatures down to 7 K. The Journal of Chemical Physics, 109(10), 3882–3897.

49. Janssen, L. M., Groenenboom, G. C., Van, D. A. A., Żuchowski, P. S., & Podeszwa, R., (2009). Ab initio potential energy surfaces for NH (Σ 3−)–NH (Σ 3−) with analytical long range. The Journal of Chemical Physics, 131(22), 224314.

50. Kirste, M., Wang, X., Schewe, H. C., Meijer, G., Liu, K., Van, D. A. A., & Van De, M. S. Y., (2012). Quantum-state resolved bimolecular collisions of velocity-controlled OH with NO radicals. Science, 338(6110), 1060–1063.

51. Landau, A., & Moiseyev, N., (2016). Molecular resonances by removing complex absorbing potentials via padé; application to CO^- and N_2^-. The Journal of Chemical Physics, 145(16), 164111.

52. Lang, F., Winkler, K., Strauss, C., Grimm, R., & Denschlag, J. H., (2008). Ultracold triplet molecules in the rovibrational ground state. Physical Review Letters, 101(13), 133005.

53. Lara, M., Chefdeville, S., Hickson, K. M., Bergeat, A., Naulin, C., Launay, J. M., & Costes, M., (2012). Dynamics of the S (D 2 1)+ HD (j= 0) reaction at collision energies approaching the cold regime: A stringent test for theory. Physical Review Letters, 109(13), 133201.

54. Lau, J. A., Toennies, J. P., & Tang, K. T., (2016). An accurate potential model for the a3Σu+ state of the alkali dimers Na_2, K_2, Rb_2, and Cs_2. The Journal of Chemical Physics, 145(19), 194308.

55. Lavert-Ofir, E., Shagam, Y., Henson, A. B., & Gersten, S., (2014). In: Los, J. K., Zuchowski, P. S., Narevicius, J., & Narevicius, E., (eds.), Nature Chem. (Vol. 6, p. 332).

56. Lee, Y. T., McDonald, J. D., LeBreton, P. R., & Herschbach, D. R., (1969). Molecular beam reactive scattering apparatus with electron bombardment detector. Review of Scientific Instruments, 40(11), 1402–1408.

57. Li, J., Jiang, B., & Guo, H., (2013). Permutation invariant polynomial neural network approach to fitting potential energy surfaces. II. Four-atom systems. The Journal of Chemical Physics, 139(20), 204103.

58. Li, Z., Apkarian, V. A., & Harding, L. B., (1997). A theoretical study of solid hydrogens doped with atomic oxygen. The Journal of Chemical Physics, 106(3), 942–953.

59. Liu, K., (2001). Crossed-beam studies of neutral reactions: State-specific differential cross-sections. Annual Review of Physical Chemistry, 52(1), 139–164.

60. Liu, K., (2016). Vibrational control of bimolecular reactions with methane by mode, bond, and stereoselectivity. Annual Review of Physical Chemistry, 67, 91–111.

61. Liu, R., Wang, F., Jiang, B., Czakó, G., Yang, M., Liu, K., & Guo, H., (2014). Rotational mode specificity in the Cl+ $_{CHD3}$→ HCl+ CD_3 reaction. The Journal of Chemical Physics, 141(7), 074310.

62. Liu, Y., & Luo, L., (2021). Molecular collisions: From near-cold to ultra-cold. Frontiers of Physics, 16(4), 1–38.

63. Liu, Y., Vashishta, M., Djuricanin, P., Zhou, S., Zhong, W., Mittertreiner, T., & Momose, T., (2017). Magnetic trapping of cold methyl radicals. Physical Review Letters, 118(9), 093201.

64. Loreau, J., & Van, D. A. A., (2015). Scattering of NH_3 and ND_3 with rare gas atoms at low collision energy. The Journal of Chemical Physics, 143(18), 184303.

65. Makrides, C., Hazra, J., Pradhan, G. B., Petrov, A., Kendrick, B. K., González-Lezana, T., & Kotochigova, S., (2015). Ultracold chemistry with alkali-metal-rare-earth molecules. Physical Review A, 91(1), 012708.

66. Maussang, K., Egorov, D., Helton, J. S., Nguyen, S. V., & Doyle, J. M., (2005). Zeeman relaxation of CaF in low-temperature collisions with Helium. Physical Review Letters, 94(12), 123002.

67. Micheli, A., Brennen, G. K., & Zoller, P., (2006). A toolbox for lattice-spin models with polar molecules. Nature Physics, 2(5), 341–347.

68. Naulin, C., & Costes, M., (2014). Experimental search for scattering resonances in near cold molecular collisions. International Reviews in Physical Chemistry, 33(4), 427–446.

69. Ni, K. K., Ospelkaus, S., De Miranda, M. H. G., Pe'Er, A., Neyenhuis, B., Zirbel, J. J., & Ye, J., (2008). A high phase-space-density gas of polar molecules. Science, 322(5899), 231–235.

70. Perreault, W. E., Mukherjee, N., & Zare, R. N., (2017). Quantum control of molecular collisions at 1 kelvin. Science, 358(6361), 356–359.

71. Qiu, M., Ren, Z., Che, L., Dai, D., Harich, S. A., Wang, X., & Zhang, D. H., (2006). Observation of Feshbach resonances in the F+ H_2→ HF+ H reaction. Science, 311(5766), 1440–1443.

72. Rabl, P., DeMille, D., Doyle, J. M., Lukin, M. D., Schoelkopf, R. J., & Zoller, P., (2006). Hybrid quantum processors: molecular ensembles as quantum memory for solid-state circuits. Physical Review Letters, 97(3), 033003.

73. Raghavachari, K., Trucks, G. W., Pople, J. A., & Head-Gordon, M., (1989). A fifth-order perturbation comparison of electron correlation theories. Chemical Physics Letters, 157(6), 479–483.
74. Rogers, S., Wang, D., Kuppermann, A., & Walch, S., (2000). Chemically accurate ab initio potential energy surfaces for the lowest 3A ‘and 3A “electronically adiabatic states of O(^{3}P) + H_2. The Journal of Physical Chemistry A, 104(11), 2308–2325.
75. Rowe, B. R., Dupeyrat, G., Marquette, J. B., & Gaucherel, P., (1984). Study of the reactions N+ 2+ $2N_2$→ N+ 4+ N_2 and O+ 2+ $2O_2$→ O+ 4+ O_2 from 20 to 160 K by the CRESU technique. The Journal of Chemical Physics, 80(10), 4915–4921.
76. Sachs, E. S., Hinze, J., & Sabelli, N. H., (1975). Frozen core approximation, a pseudopotential method tested on six states of NaH. The Journal of Chemical Physics, 62(9), 3393–3398.
77. Sahai, R., & Nyman, L. Å., (1997). The boomerang nebula: The coldest region of the universe? The Astrophysical Journal Letters, 487(2), L155.
78. Sawyer, B. C., Lev, B. L., Hudson, E. R., Stuhl, B. K., Lara, M., Bohn, J. L., & Ye, J., (2007). Magnetoelectrostatic trapping of ground-state OH molecules. Physical Review Letters, 98(25), 253002.
79. Sawyer, B. C., Stuhl, B. K., Wang, D., Yeo, M., & Ye, J., (2008). Molecular beam collisions with a magnetically trapped target. Physical Review Letters, 101(20), 203203.
80. Sawyer, B. C., Stuhl, B. K., Yeo, M., Tscherbul, T. V., Hummon, M. T., & Xia, Y., (2011). Los, J. K., Patterson, D., Doyle, J. M., & Ye, J., (eds.), Phys. Chem. Chem. Phys., 13(19059), 1008–5127.
81. Schnieder, L., Seekamp-Rahn, K., Borkowski, J., Wrede, E., Welge, K. H., Aoiz, F. J., & Herrero, V. J., (1995). In: Sa ez, R. B. V., & Wyatt, R. E., (eds.), Science (Vol. 269, p. 207).
82. Shagam, Y., Klein, A., Skomorowski, W., Yun, R., Averbukh, V., Koch, C. P., & Narevicius, E., (2015). Molecular hydrogen interacts more strongly when rotationally excited at low temperatures leading to faster reactions. Nature Chemistry, 7(11), 921.
83. Shuman, E. S., Barry, J. F., & DeMille, D., (2010). Laser cooling of a diatomic molecule. Nature, 467(7317), 820–823.
84. Shuman, E. S., Barry, J. F., Glenn, D. R., & DeMille, D., (2009). Radiative force from optical cycling on a diatomic molecule. Physical Review Letters, 103(22), 223001.

85. Skodje, R. T., Skouteris, D., Manolopoulos, D. E., Lee, S. H., Dong, F., & Liu, K., (2000). Resonance-Mediated Chemical Reaction: F+ H D→ H F+ D. Physical Review Letters, 85(6), 1206.
86. Sorbie, K. S., & Murrell, J. N., (1975). Analytical potentials for triatomic molecules from spectroscopic data. Molecular Physics, 29(5), 1387–1407.
87. Sowokinos, J. R., (2001). Biochemical and molecular control of cold-induced sweetening in potatoes. American Journal of Potato Research, 78(3), 221–236.
88. Thomas, N. R., Kjærgaard, N., Julienne, P. S., & Wilson, A. C., (2004). Imaging of s and d partial-wave interference in quantum scattering of identical bosonic atoms. Physical Review Letters, 93(17), 173201.
89. Thorsheim, H. R., Weiner, J., & Julienne, P. S., (1987). Laser-induced photoassociation of ultracold sodium atoms. Physical Review Letters, 58(23), 2420.
90. Van De, M. S. Y., & Meijer, G., (2009). Collision experiments with Stark-decelerated beams. Faraday Discussions, 142, 113–126.
91. Van, D. P. A. P. P., & Bethlem, H. L., (2018). A detailed account of the measurements of cold collisions in a molecular synchrotron. EPJ Techniques and Instrumentation, 5, 1–27.
92. Vogels, S. N., Karman, T., Kłos, J., Besemer, M., Onvlee, J., van der Avoird, A., & van de Meerakker, S. Y. (2018). Scattering resonances in bimolecular collisions between NO radicals and H 2 challenge the theoretical gold standard. Nature chemistry, 10(4), 435-440.
93. Vogels, S. N., Onvlee, J., Chefdeville, S., Van, D. A. A., Groenenboom, G. C., & Van De, M. S. Y., (2015). Imaging resonances in low-energy NO-He inelastic collisions. Science, 350(6262), 787–790.
94. Wallis, A. O., Longdon, E. J., Żuchowski, P. S., & Hutson, J. M., (2011). The prospects of sympathetic cooling of NH molecules with Li atoms. The European Physical Journal D, 65(1), 151–160.
95. Wang, F., Lin, J. S., & Liu, K., (2011). Steric control of the reaction of CH stretch-excited CHD_3 with chlorine atom. Science, 331(6019), 900–903.
96. Wang, T. T., Heo, M. S., Rvachov, T. M., Cotta, D. A., & Ketterle, W., (2013). Deviation from universality in collisions of ultracold Li 2 6 molecules. Physical Review Letters, 110(17), 173203.

97. Wei, Q., Lyuksyutov, I., & Herschbach, D., (2012). Merged-beams for slow molecular collision experiments. The Journal of Chemical Physics, 137(5), 054202.
98. Weiner, J., Bagnato, V. S., Zilio, S., & Julienne, P. S., (1999). Experiments and theory in cold and ultracold collisions. Reviews of Modern Physics, 71(1), 1.
99. Weinstein, J. D., DeCarvalho, R., Guillet, T., Friedrich, B., & Doyle, J. M., (1998). Magnetic trapping of calcium monohydride molecules at milli-Kelvin temperatures. Nature, 395(6698), 148–150.
100. Werner, H. J., & Knowles, P. J., (1985). A second order multiconfiguration SCF procedure with optimum convergence. The Journal of Chemical Physics, 82(11), 5053–5063.
101. Werner, H. J., Knowles, P. J., Knizia, G., Manby, F. R., & Schütz, M., (2012). Molpro: A general-purpose quantum chemistry program package. Wiley Interdisciplinary Reviews: Computational Molecular Science, 2(2), 242–253.
102. Willey, D. R., Crownover, R. L., Bittner, D. N., & De Lucia, F. C., (1988). Very low temperature spectroscopy: The pressure broadening coefficients for CO–He between 4.3 and 1.7 K. The Journal of Chemical Physics, 89(4), 1923–1928.
103. Wu, X., Gantner, T., Koller, M., Zeppenfeld, M., Chervenkov, S., & Rempe, G., (2017). A cryofuge for cold-collision experiments with slow polar molecules. Science, 358(6363), 645–648.
104. Wynar, R., Freeland, R. S., Han, D. J., Ryu, C., & Heinzen, D. J., (2000). Molecules in a Bose-Einstein condensate. Science, 287(5455), 1016–1019.
105. Xiao, C., Xu, X., Liu, S., Wang, T., Dong, W., Yang, T., & Yang, X., (2011). Experimental and theoretical differential cross sections for a four-atom reaction: HD+ OH→ H_2O+ D. Science, 333(6041), 440–442.
106. Xu, K., Mukaiyama, T., Abo-Shaeer, J. R., Chin, J. K., Miller, D. E., & Ketterle, W., (2003). Formation of quantum-degenerate sodium molecules. Physical Review Letters, 91(21), 210402.
107. Yang, X., & Zhang, D. H., (2008). Dynamical resonances in the fluorine atom reaction with the hydrogen molecule. Accounts of Chemical Research, 41(8), 981–989.

108. Yang, X., (2007). State-to-state dynamics of elementary bimolecular reactions. Annu. Rev. Phys. Chem., 58, 433–459.

109. Zeppenfeld, M., Englert, B. G., Glöckner, R., Prehn, A., Mielenz, M., Sommer, C., & Rempe, G., (2012). Sisyphus cooling of electrically trapped polyatomic molecules. Nature, 491(7425), 570–573.

110. Zieger, P. C., Eyles, C. J., Van De, M. S. Y., Van, R. A. J., Bethlem, H. L., & Meijer, G., (2013). A forty-segment molecular synchrotron. Zeitschrift für Physikalische Chemie, 227(11), 1605–1645.

111. Zirbel, J. J., Ni, K. K., Ospelkaus, S., D'Incao, J. P., Wieman, C. E., Ye, J., & Jin, D. S., (2008). Collisional stability of fermionic Feshbach molecules. Physical Review Letters, 100(14), 143201.

Chapter

4

Interaction Among Cold Atoms and Molecules

CONTENTS

4.1. INTRODUCTION

Examining and understanding the fundamental interaction among particles has continually been a motivating force in scientific advance. With adequate knowledge of the primary mechanisms, one could tailor methods to directly regulate this interaction down to the solo particle level, permitting experimental entrance to new phenomena and allowing the design of complicated physical systems of interest. Instances of these achievements could be found in the latest scientific history within the success stories of neutral atoms, and laser-cooled charged, occasioning four Physics Nobel prizes (1989, 1997, 2001, and 2012). Restraining charged particles with magnetic and electric fields dates back to the 1950s. With the initiation of laser cooling some 20 years future, the formation of single trapped ions became an actuality providing the road for complicated experiments (Dehmelt, 1975; Wineland & Itano, 1981). Currently, ions are regularly cooled to the quantum mechanical ground state of their limitation in laboratories all around the globe. However, multi-qubit quantum computation with trapped ions has been shown creating these systems the typical instance of completely controllable quantum systems. Trapping of neutral atoms has constantly been a more complicated topic as applicable exterior forces are feebler likened to the electric force on charged particles causing in shallow traps. Laser cooling methods were crucial in decreasing the kinetic energies adequately and finally permitted for the gathering of cold atoms in magnetic optical or magneto-optical traps (MOTs) (Raab et al., 1987; Miller et al., 1993). This initiating point allowed the development of evaporative cooling methods and finally made Bose-Einstein concentration in dilute atomic gases a realism Initiating with direct tunability of the inter-particle interaction through Fesh bach characters tremendous development has been done over the subsequent years (Weitenberg et al., 2011). Currently, ultracold quantum gases permit preparation and read-out at the single-atom level making them the top platform to study the dynamics of many-body quantum systems (Inouye et al., 1998).

The overall development of laser cooling, trapping, and operating both ultracold neutral atoms and charged ions has permitted exceptional experimental control of external and internal degrees of freedom (DOF) on the multi-particle and single-particle levels. At the start of the 21st century, suggestions were brought up to examine combined ultracold atom-ion systems as the resemblance in technology and utilized methods provide very good starting circumstances for experiments. Examination of cold chemistry in charged gases at very little collision energies was however one of the subjects of interest (Côté, 2000; Smith et al., 2003). Getting these collision energies were erroneously straightforward as ultracold atoms should considerately cool ions through elastic collisions (Krych et al., 2014). Utilizing previously established tools from single ion experiments and ultracold atoms, one could examine atom-ion physics in regimes that weren't available before.

The novel research field was also determined through the promise of having a dissimilar inter-particle interaction likened to what has been used so far with ions or neutrals only. With an underlying long-range and strong polarization potential among an ion and atom suggestions were brought up to utilize hybrid atom-ion structures as a testbed to study intensely interrelated many-body effects (Cote et al., 2002; Massignan et al., 2005) or for quantum information and simulation drives. A quantum mechanical theory of ultracold atom ion crash was formed, forecasting Fesh bach resonances (Idziaszek et al., 2009; Gao, 2010). These theoretical insights stimulated continuing experimental work to ultracold atom-ion systems. Feshbach resonances would permit a direct tunability of the inter-particle interaction, as soon as sufficient low collision energies have been comprehended experimentally (Kais & Shi, 2000; Idziaszek et al., 2007).

4.2. THE POLARIZATION POTENTIAL

The electric field of a charge provides growth to the long-range $r{-}^{1}$ and strong Coulomb potential, describing the repulsive interaction among ions with a similar charge. The interaction among polarizable neutral atoms is regulated through induced dipole moments which consequence in a short-range $r{-}^{6}$ and generally attractive Van-der Waals potential. The particles' inner spin structure is only distinguished if they could approach each other sufficiently close for their spins to interrelate. With diverse range features of the two potentials, spin orientations become inappropriate for interacting ions however harmful when neutral atoms strike (Monroe et al., 1995). The atom-ion polarization potential signifies an intermediate case permitting for both an internal spin state reliance and long-range interactions (Anderson et al., 1995; Davis et al., 1995).

Having ion and atom separated through a distance r, the ion's electric field $E(r) = \frac{e}{4\pi\epsilon_0 r^2}$, with the vacuum permittivity $_0$ and elementary charge e, brings a dipole $p(r) = 4\pi \in_0 \alpha E(r)$ given through the atoms polarizability α. This induced dipole in the ion's electric field takes to the polarization potential with the coefficient $C_4 = \frac{\alpha e^2}{4\pi\epsilon_0}$.

$$V(r) = -\frac{1}{2}p(r)E(r) = -\frac{1}{2}\frac{C_4}{r^4},$$

For the explanation of a collision procedure in a radial symmetric potential, it is appropriate to employ spherical manages in the center-of-mass reference frame. Additionally to the interaction potential, centrifugal energy seems, introducing an efficient radial potential that relies on the angular momentum. Taking quantum mechanical dispersion, this angular momentum is quantized, providing rise to precise potentials for every angular momentum quantum number *l*. This method is recognized as the partial wave expansion, permitting the total collision cross-section to be divided into the independent contributions of every partial wave with its angular momentum *l*. By equating the interaction potential $\frac{C_4}{2r^4}$ with the centrifugal energy, $\frac{2\bar{h}^2}{2\mu r^2}$ one describes a length scale $R^* = \sqrt{\frac{\mu C_4}{\bar{h}^2}}$ and the related energy $E^* = \frac{\bar{h}^2}{2\mu R^{*2}}$ which are features for the interaction potential. $E*$ is associated with the lowest centrifugal barrier, indicating that at collision energy on the direction of $E*$ all partial waves however the $l = 0$ waves, the *s*-wave, is reflected off their particular centrifugal barrier. In further words, the *s*-wave entirely defines the collision procedure making a quantum mechanical treatment indispensable (Monz et al., 2016). The *ultracold* regime directly mentions *s*-wave collisions. With enhancing the number of contributing partial waves, the collision could then again be explained classically. Generally, low collision energies joined with short-range interactions and low reduced mass push to the ultracold regime. For the system of attention Ba^+-Rb, $E* = 52$nK, however for neutral Rb-Rb $E_{RbRb}* = 292\mu$K. Therefore, multiple partial waves could be appropriate for Ba^+-Rb at collision energies wherever neutral Rb-Rb is previously in the *s*-wave regime. This makes atom-ion collisions slightly peculiar as a classical description could be entirely sufficient regardless of having collision energies in the low mK range. Looking at the specific length scales, being R_{RbRb}^* 4nm *and* $R^* =$ 295 nm, one could see a difference by nearly 2 orders of magnitude. This comparison proposes that atom-ion systems could be utilized as a promising testbed for strongly interrelated many-body physics as the underlying long-range potential helps many-body interactions (Webb et al., 1995).

4.3. REALIZING (ULTRA-)COLD ATOM-ION EXPERIMENTS

Around a dozen laboratories globally are executing cold atom-ion experiments and some more are being set up. In contrast with earlier experiments with drift tube or beam/target both charged and neutral particles are trapped at a similar time, providing very low collision energies. Generally, all cold atom-ion experiments initiate with laser-cooled atoms, most also utilize laser-cooled ions. A huge group of laser cool-able species is signified by alkali-metal and alkali-earth elements. Generally, the earlier is selected for neutral atoms and the further for charged ions because of the simple hydrogen-like electronic configuration with a solo valence electron. These initial two groups of the periodic table only permit for a plethora of species atom-ion mixtures that could be experimentally examined. In most of the experiments, a mixture of $^{138}Ba^{+}$ and ^{87}Rb was selected (Makarov et al., 2003). This system permits for many elastic collisions earlier to an inelastic collision takes place because of a strong suppression of charge transfer. Seeing at the proposals stated in the preceding section gaining experimental rights to the regime of ultracold atom-ion collisions looks like a very promising short-term goal. The term "ultracold" normally defines a range of energies where collisions could be explained quantum mechanically and more precisely through a single partial wave, the s-wave. Numerous groups are working towards atom-ion s-wave smashes, which so far have not yet been shown. The necessary low collision energy, because of the small characteristic energy $E*$, is just one of the causes why this curved out to be non-trivial. Numerous obstacles have been recognized by now that have to be overwhelmed, which we will explain in this section after sketching how to realize a hybrid atom-ion experiment (Goold et al., 2010; Casteels et al., 2011).

4.3.1. Hybrid atom-Ion Trapping

There is a multitude of diverse trap types being utilized as workhorses in both communities of charged ions and ultracold neutral atoms. The related technology is currently well recognized and it could principle be directly used for atom-ion experiments. The only remaining contest when drafting a hybrid atom-ion trap, is the special mixture of the ion and the atom trap, permitting for coinciding trap volumes of both traps. In the succeeding, we would concisely describe diverse viable atom and ion trapping techniques. This list is by no means comprehensive as it only reflects the so far used traps in atom-ion experiments (Idziaszek et al., 2011).

4.3.1.1. Atom Traps

Presently, three distinct kinds of atom traps are in usage. After their respective up and downsides, they differ in their attainable atomic densities and temperatures, generally set through the experimentally appropriate cooling mechanisms in these traps.

4.3.1.2. Ion Traps

The history of ion trapping surpassed that for neutral trapping through several decades. Currently, single-ion trapping has proceeded routinely in numerous laboratories around the world. Out of the two major trap kinds, only the Paul trap is being utilized for atom-ion experiments. The penning trap doesn't look tempting mainly due to the huge magnetic fields that are needed for steady ion trapping. Producing these field strengths needs a trap design that generally limits optical access greatly and thus the compatibility with atom dipole traps. In penning traps, the laser cooling of ions is also non-trivial because of the involved motional modes (Thompson & Papadimitriou, 2000). Further considerations have to be done when trying to present ultracold atoms into the huge magnetic fields of a penning trap.

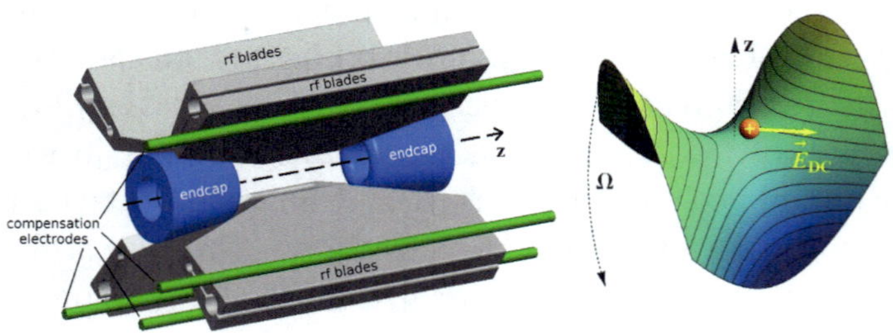

Figure 4.1. Left: Linear Paul trap utilized in atom-ion apparatus in Ulm. Right: The consequential radial quadrupole potential wavers at a frequency Ω and gives the rose to a trapped ion trajectory.

Note: Two-dimensional limitation in the radial plane is produced through applying a sinusoidal voltage at a frequency Ω to the RF blades with a phase shift of π among adjoining blades. Static voltages are used to the endcaps and give axial limitation in the third dimension. Two sets of compensation electrodes could be utilized to create oscillating electric forces or additional static in the radial plane. The RF node could be found alongside the z-axis of a linear Paul trap. Electrical stray fields $E_{\sim DC}$ move the trap center away from the RF node, persuading additional micromotion.

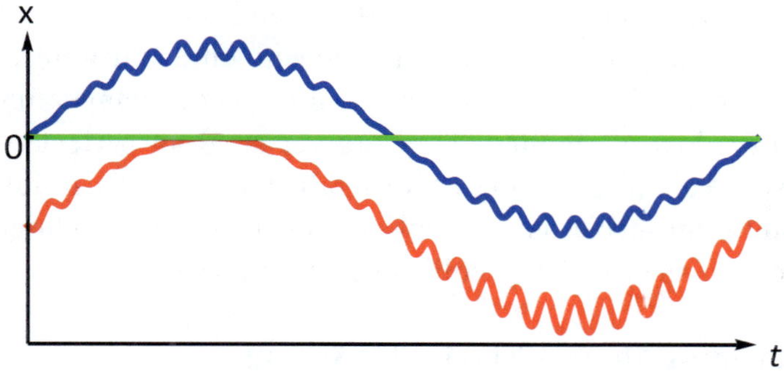

Figure 4.2. Upper trajectory: The ion trajectory could be disintegrated in a slow harmonic oscillation altered with a fast micromotion. The micromotion is proportional to the local RF field amplitude, which rises linearly with distance from the RF node. Lower trajectory: Extra static electrical fields produce a constant force, moving the trap center away from the RF field node, bringing additional micromotion.

4.3.2. Sympathetic Cooling

Sympathetic cooling in principle only needs a buffer gas functioning as a refrigerator. Utilizing a refrigerator for cooling can't merely be conceptually simpler likened to more complicated laser cooling methods, like the stated ground-state cooling however also applicable to an extensive range of species. This is due to laser cooling is restricted to particles where closed cycling transitions are accessible. On the other hand, the concept of sympathetic cooling is appropriate even without precise knowledge of accessible internal states (Krasniqi et al., 2010; Bissbort et al., 2013). The momentum transfer in elastic smashes among cold buffer gas atoms and the hot target assist a redistribution of kinetic energy till an equilibrium temperature is got. The only requisite is that many elastic collisions could take place earlier to an inelastic collision happens. Inelastic collisions could introduce heating through the discharged energy when internal states of the particles are altered, i.e., spin flips, and even alter the species in a chemical reaction. Charge transfer of the type $A^+ + B \rightarrow A + B^+$ or molecular formation $A^+ + B \rightarrow AB^+$ have been the most examined inelastic procedures in cold atom-ion systems till yet. Generally, chemically inert buffer gases, like noble gases, are utilized for sympathetic cooling to overwhelm inelastic collisions. A classic instance is Helium buffer gas cooling down to the K range which could be stretched down to tens of mK in a helium dilution cryostat. Utilizing noble gases for trials in the emerging cold atom-ion field though is not a feasible option yet as the attainable temperatures can't compete with ultracold atoms. Ultracold atomic gases create the perfect buffer gas in respect of temperature as they could be in the low nK regime. They are yet far from perfect concerning reactivity as one classically works with alkali atoms. Thus, careful selection of the ion and atom species is essential as sympathetic cooling might only be feasible if reactions or inelastic collisions occur very rarely.

4.3.3. Micromotion and Elastic Collisions

Thermalization of a solo ion in interaction with an ultracold atomic bath would be the preferred situation to reach ultracold collision energies. This though doesn't happen in current generation hybrid atom-ion experiments. The only cause for this phenomenon is the dynamic ion trapping in a Paul trap. The driven micromotion could transmit energy into the secular motion through elastic atom-ion collisions. This could be understood as a phased movement of the ion's secular oscillation, relying not merely on the momentum exchange among ion and atom however also on the precise phase of the micromotion at the time of the atom-ion smash. A decrease or increase of the secular oscillation amplitude is probable in such a situation (see Figure 4.3). The non-conservative trapping perspective in combination with collisions could efficiently introduce extra energy into the system, also recognized as micromotion heating. Now, the subsequent equilibrium energies could be far beyond the temperatures of the ultracold atomic gas utilized for sympathetic cooling (Anderson et al., 1995; Davis et al., 1995)

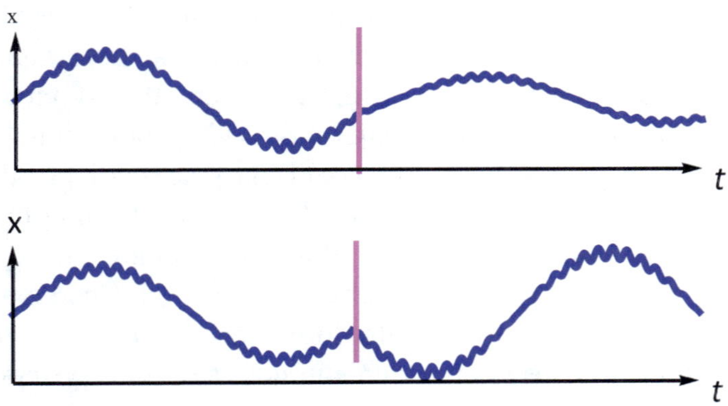

Figure 4.3. One dimensional ion trajectory in a Paul trap over the period, striking with an atom at rest (dashed line). The elastic collision would transmit momentum from the ion onto the atom, decreasing the ion's instant velocity.

Note: As the micromotion is determined, its micromotion velocity would be restored after the collision, efficiently pairing the secular oscillation with the micromotion. In this method, the last ion velocity is not only determined through momentum transfer through the collision. Relying on the precise micromotion phase at the time of the crash the secular energy could be reduced (upper panel) or enhanced (lower panel).

4.3.4. Inelastic Processes

A comprehensive investigation of accessible inelastic procedures in cold atom-ion collisions is significant for the future growth of the entire field. The instance of sympathetic cooling previously showcases this as it could only be possible (comprising down to s-wave collision energies) if inelastic procedures don't happen frequently. Looking at a wider context, this relates to any type of atom-ion interaction that is facilitated through elastic collisions. This is obvious when looking at many-body offers like the polaron formation or mesoscopic molecular ion (Cote et al., 2002; Casteels et al., 2011). It is highly implausible that the made many-body state would be the accurate ground state of the mutual system as only the long-range interaction is being deliberated in the proposals. Any type of transition to a lower-lying state would discharge energy, possibly destroying the many-body system. This indicates that the time scales of the accessible inelastic channels would be fixed a normal time limit for experiments within these systems.

4.3.4.1. Two-Body Inelastic Collisions

In common, a collision is called inelastic when not the only momentum is being substituted; however also an alteration of the internal states of the particles happens. A prominent instance, known from ultracold atoms, are spin flips or generally, hyperfine transitions. If spin altered during a collision among the particles is probable, for example, the spin configuration is not extended, the related energy could be retrieved, providing harm and/or heating mechanism. This procedure can also happen in cold atom-ion collisions and has been examined with $^{174}Yb^{+}$-^{87}Rb (Ratschbacher et al., 2013).

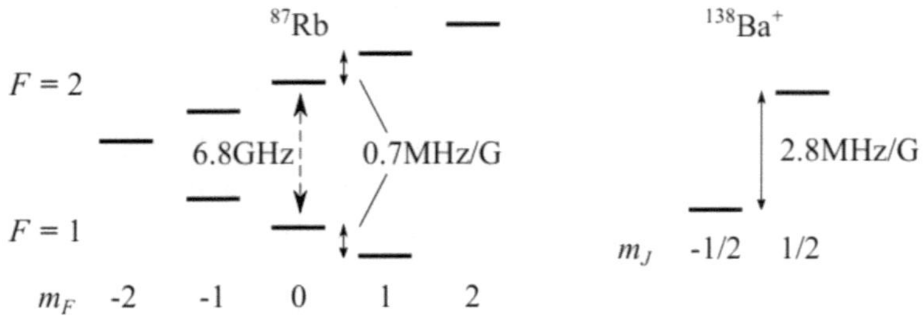

Figure 4.4. Illustrated are the ground states of 87Rb and 138Ba+. This comprises the Hyperfine structure for 87Rb and the Zeeman splittings of both systems.

Note: Spin exchange collisions among the two systems could potentially discharge energy. Making spin stretched configurations could overwhelm this procedure if no spin relaxation happens. Making in the lowest accessible states of both structures should in principle deny this kind of inelastic collisions.

The related level structure of the $^{138}Ba^{+}$-^{87}Rb system is the same as $^{174}Yb^{+}$-^{87}Rb (see Figure 4.4), proposing that the perceived spin-exchange and relaxation should happen here as well. Formation of ^{87}Rb atoms in the $F = 2$ hyperfine state permits for collisional transitions to the lower-lying $F = 1$ state, which could release $h \times 6.8$ GHz = $k_B \times 325$ mK of energy. However, when working with the $F = 1$ manifold one has to prudently select the m_F Zeeman substates. As a magnetic preference field of a few Gauss is generally always existing in experiments, comprising a Zeeman-splitting of 0.7 MHz/G on the m_F states. For a general offset field of 2 G and spin-polarized ^{87}Rb atoms in $F = 1$ $m_F = -1$ (a magnetic trappable condition), this corresponds to $h \times 2.8$ MHz = $k_B \times 130$ μK of energy that could potentially be retrieved through collisions with the ion. A similar line of arguments relates to the state creation of the ion where the Zeeman-splitting of the two ground states comes into work. Denying this kind of inelastic collisions is possible through preparing both systems in their relevant entire lowest energy hyperfine and/or Zeeman state.

Chemical reactions, however, could still happen since the combined atom-ion system doesn't have to be in its electronic ground state however, if both systems on their specific area in their relevant ground states. The Ba^{+}-Rb system is an instance of this behavior. Here, the originally prepared Ba^{+}-Rb is actively roughly 1eV above the electronic configuration of a neutral Ba and an ionized Rb^{+} (see Figure 4.5) (Thompson & Papadimitriou, 2000). Molecule formation and charge transfer are a result of the system's transition among these two electronic configurations. In this precise system, the transition is only probable through the discharge of a photon, as in radiative charge transfer, which decreases the related probability for these procedures and permits for 10^5 elastic collisions earlier one charge transfer collision happens (Krych et al., 2011).

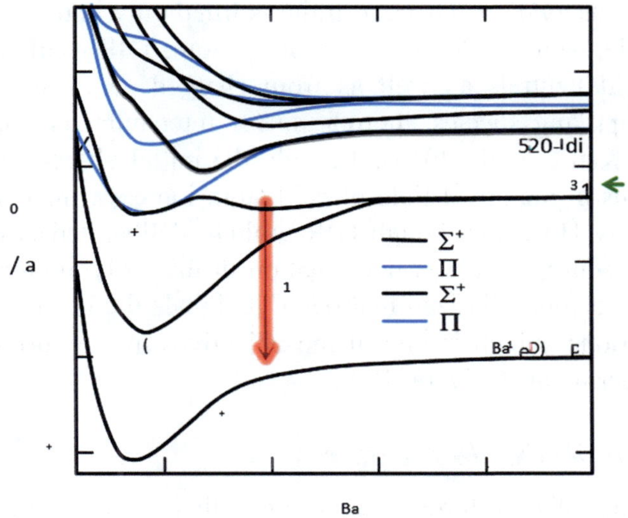

Figure 4.5. Molecular potential curves for the Ba+-Rb system displaying the collisional entrance channel (green arrow) together with the lower-lying ground state and several excited states.

Note:Transitions to the lesser potential, asymptotically linked to the electronic configuration Ba-Rb^+, are only probable through photon emission. The plethora of low-lying higher potentials could become appropriate for light-assisted collisions. Now, only the lowest asymptotes are displayed. However, long-wavelength infrared dipole traps at 1550 nm or 1064 nm could in principle become appropriate as they could attach the entrance channel to the higher repulsive potential curves. Potentials took from (Leimkuhler et al., 2013).

4.3.4.2. Many-Body Inelastic Collisions

One of the basic reasons to be attentive in cold atom-ion systems is the primary r^{-4} polarization potential, which assure to be long-range and strong likened to the r^{-6} Van-der-Waals interface in neutral atoms. The strong interaction assists many-body phenomena, making a theoretical and experimental examination of atom-ion systems gorgeous. However, a many-body system intrinsically introduces bound states at lesser energies, potentially permitting novel inelastic procedures. Here, a prime instance is three-body recombination (TBR) that explains molecule creation in a ternary collision. Binary collisions on their own can't enable molecule creation as a single product particle would break up the laws of energy and momentum conservation. This constraint is lifted only in the occurrence of a third particle, be it an additional photon or atom. TBR itself is recognized from ultracold neutrals as well as from charged gases and plasmas in general, comprising species appropriate for interstellar chemistry (Plašil et al., 2012; Krstić et al., 2013). Though, the initial observation of three-body reactions in the novel field of cold atom-ion experiments was stated only very lately. However, though TBR could be deliberated as an undesired inelastic procedure, as it was the situation in the early attempts to Bose-Einstein reduce spin-polarized hydrogen, it signals the inception of many-body interactions, which is one of the objectives in the current day cold atom-ion experiments (Silvera, 1992).

4.3.4.3. Light Assisted Processes

The occurrence of light through atom-ion collisions has to be deliberated as well when conniving experiments. This goes past the evident issues when working with a continually laser-cooled atom-ion system, that uses a Coulomb crystal and/or MOT. In both these circumstances, near-resonant cooling light has to be existing throughout the experiments. This permits for photo-association and crashes in excited states which introduce massively diverse reaction dynamics (Sikorsky et al., 2017).

However, even off-resonant light could affect the collision dynamics. Laser light that is off-resonant for both classes on their own, might yet drive alterations to exciting potential curves of the joint system. A usual situation is the usage of an optical dipole trap at 1550 nm or 1064 nm to stock the ultracold atoms. Similar to the requirement of persistent laser cooling in an MOT this light is required for trapping and is thus always present. Looking at the Ba^+-Rb system, there are numerous low-lying asymptotes overhead the entrance channel, which could probably be retrieved with infrared photons (Figure 4.5). Here the usage of 1064nm light has not turned out to be challenging so far, however, its role should be deliberated as soon as many-body collisions come into play. The subsequent excited potential energy surfaces (PESs) of Ba^+-N Rb ($N \geq 2$) might in principle backed transitions with 1064nm photons (Esry et al., 1999).

4.4. THE ATOM-ION TRAP IN ULM

The experimental arrangement in Ulm is one of the initial two-hybrid atom-ion traps that could prepare debased atomic samples. Being in working since 2009, a detailed explanation of the machine and operation principle previously exists (Zhang & Tanaka, 1998; Schmid et al., 2012). Therefore, we would limit ourselves only to concisely describing the core of the entire lab, the vacuum system that comprises the hybrid atom-ion trap.

The vacuum system comprises three linked chambers named after their relevant roles (lower left in Figure 4.6). Rb atoms are stuck in an MOT loaded from background vapor then magnetically conveyed into a chamber with sufficient low pressure for evaporative cooling down to Bose-Einstein condensation (BEC). Later, an optical transport into the middle of the Paul trap, Rb atoms are burdened in a crossed dipole trap (upper right in Figure 4.6). The Paul trap could store single ions, for most of the present experiments both Ba^+ and Rb^+ ions are utilized routinely. During the last preparation phase, we overlap the two trap centers and submerge the ion into a cloud of ultracold atoms (lower right in Figure 4.6) (Schmid et al., 2010).

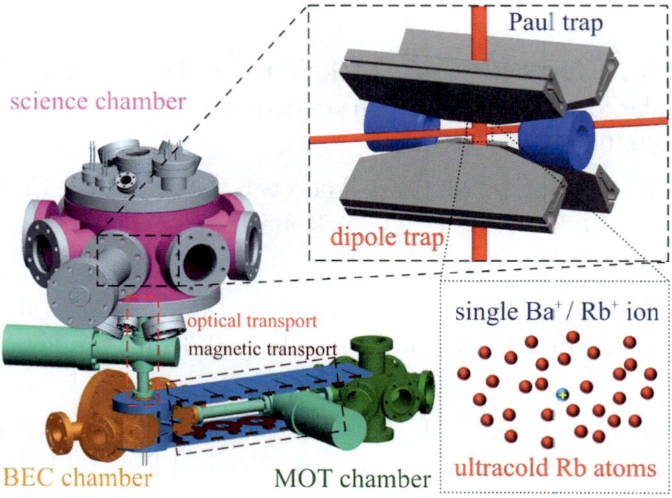

Figure 4.6. Lower Left the vacuum system comprising of a BEC, MOT, and science chamber together with the optical and magnetic transport sections for the Rb atoms. Upper right the heart of the experimental system, the combined atom-ion trap. Two laser beams at 1064 nm create a crossed dipole trap inside the trapping volume of a linear Paul trap. Lower right spatially overlying trap centers permit the engagement of single ions (typically Ba+ or Rb+) into ultra-cold Rb atom clouds.

REFERENCES

1. Allcock, D. T. C., Guidoni, L., Harty, T. P., Ballance, C. J., Blain, M. G., Steane, A. M., & Lucas, D. M., (2011). Reduction of heating rate in a microfabricated ion trap by pulsed-laser cleaning. New Journal of Physics, 13(12), 123023.
2. Anderson, M. H., Ensher, J. R., Matthews, M. R., Wieman, C. E., & Cornell, E. A., (1995). Observation of Bose-Einstein condensation in a dilute atomic vapor. Science, 269(5221), 198–201.
3. Bissbort, U., Cocks, D., Negretti, A., Idziaszek, Z., Calarco, T., Schmidt-Kaler, F., & Gerritsma, R., (2013). Emulating solid-state physics with a hybrid system of ultracold ions and atoms. Physical Review Letters, 111(8), 080501.
4. Casteels, W., Tempere, J., & Devreese, J. T., (2011). Polaronic properties of an ion in a Bose-Einstein condensate in the strong-coupling limit. Journal of Low Temperature Physics, 162(3), 266–273.
5. Côté, R., (2000). From classical mobility to hopping conductivity: Charge hopping in an ultracold gas. Physical Review Letters, 85(25), 5316.
6. Cote, R., Kharchenko, V., & Lukin, M. D., (2002). Mesoscopic molecular ions in Bose-Einstein condensates. Physical Review Letters, 89(9), 093001.
7. Davis, K. B., Mewes, M. O., & Andrews, M. R., (1995). In: Van, D. N. J., Durfee, D. S., Kurn, D. M., & Ketterle, W., (eds.), Phys. Rev. Lett. (Vol. 75, p 3969).
8. Dehmelt, H. G., (1975). Proposed 10^< 14> v/Δv laser fluorescence spectroscopy on T1^+ monoion oscillator II. Bull. Am. Phys. Soc., 20, 60.
9. Esry, B. D., Greene, C. H., & Burke, Jr. J. P., (1999). Recombination of three atoms in the ultracold limit. Physical Review Letters, 83(9), 1751.
10. Gao, B., (2010). Universal properties in ultracold ion-atom interactions. Physical Review Letters, 104(21), 213201.
11. Goold, J., Doerk, H., Idziaszek, Z., Calarco, T., & Busch, T., (2010). Ion-induced density bubble in a strongly correlated one-dimensional gas. Physical Review A, 81(4), 041601.

12. Härter, A., Krükow, A., Brunner, A., & Denschlag, J. H., (2014). Long-term drifts of stray electric fields in a Paul trap. Applied Physics B, 114(1), 275–281.
13. Huang, Y., Qi, J., Pechkis, H. K., Wang, D., Eyler, E. E., Gould, P. L., & Stwalley, W. C., (2006). Formation, detection and spectroscopy of ultracold Rb2 in the ground X1Σg+ state. Journal of Physics B: Atomic, Molecular and Optical Physics, 39(19), S857.
14. Huber, T., Lambrecht, A., Schmidt, J., Karpa, L., & Schaetz, T., (2014). A far-off-resonance optical trap for a Ba^+ ion. Nature Communications, 5(1), 1–7.
15. Idziaszek, Z., Calarco, T., & Zoller, P., (2007). Controlled collisions of a single atom and an ion guided by movable trapping potentials. Physical Review A, 76(3), 033409.
16. Idziaszek, Z., Calarco, T., Julienne, P. S., & Simoni, A., (2009). Quantum theory of ultracold atom-ion collisions. Physical Review A, 79(1), 010702.
17. Idziaszek, Z., Simoni, A., Calarco, T., & Julienne, P. S., (2011). Multichannel quantum-defect theory for ultracold atom-ion collisions. New Journal of Physics, 13(8), 083005.
18. Inouye, S., Andrews, M. R., Stenger, J., Miesner, H. J., Stamper-Kurn, D. M., & Ketterle, W., (1998). Nature, 392, 151.
19. Kais, S., & Shi, Q., (2000). Quantum criticality and stability of three-body Coulomb systems. Physical Review A, 62(6), 060502.
20. Kepple, P., & Griem, H. R., (1968). Improved Stark profile calculations for the hydrogen lines H α, H β, H γ, and H δ. Physical Review, 173(1), 317.
21. Koch, V., & Andrae, D., (2011). Static electric dipole polarizabilities for isoelectronic sequences. International Journal of Quantum Chemistry, 111(4), 891–903.
22. Krasniqi, F., Najjari, B., Strüder, L., Rolles, D., Voitkiv, A., & Ullrich, J., (2010). Imaging molecules from within: Ultrafast angström-scale structure determination of molecules via photoelectron holography using free-electron lasers. Physical Review A, 81(3), 033411.
23. Krstić, P. S., Janev, R. K., & Schultz, D. R., (2003). Three-body, diatomic association in cold hydrogen plasmas. Journal of Physics B: Atomic, Molecular and Optical Physics, 36(16), L249.

24. Krych, M., & Idziaszek, Z., (2015). Description of ion motion in a Paul trap immersed in a cold atomic gas. Physical Review A, 91(2), 023430.
25. Krych, M., Skomorowski, W., Pawłowski, F., Moszynski, R., & Idziaszek, Z., (2011). Sympathetic cooling of the Ba+ ion by collisions with ultracold Rb atoms: Theoretical prospects. Physical Review A, 83(3), 032723.
26. Labaziewicz, J., Ge, Y., Leibrandt, D. R., Wang, S. X., Shewmon, R., & Chuang, I. L., (2008). Temperature dependence of electric field noise above gold surfaces. Physical Review Letters, 101(18), 180602.
27. Lammers, J., Weimer, H., & Hammerer, K., (2016). Open-system many-body dynamics through interferometric measurements and feedback. Physical Review A, 94(5), 052120.
28. Leimkuhler, B., Margul, D. T., & Tuckerman, M. E., (2013). Stochastic, resonance-free multiple time-step algorithm for molecular dynamics with very large time steps. Molecular Physics, 111(22, 23), 3579–3594.
29. Lozeille, J., Fioretti, A., Gabbanini, C., Huang, Y., Pechkis, H. K., Wang, D., & Dulieu, O., (2006). Detection by two-photon ionization and magnetic trapping of cold Rb_2 triplet state molecules. The European Physical Journal D-Atomic, Molecular, Optical and Plasma Physics, 39(2), 261–269.
30. Makarov, O. P., Côté, R., Michels, H., & Smith, W. W., (2003). Radiative charge-transfer lifetime of the excited state of (NaCa)+. Physical Review A, 67(4), 042705.
31. Massignan, P., Pethick, C. J., & Smith, H., (2005). Static properties of positive ions in atomic Bose-Einstein condensates. Physical Review A, 71(2), 023606.
32. Miller, J. D., Cline, R. A., & Heinzen, D. J., (1993). Far-off-resonance optical trapping of atoms. Physical Review A, 47(6), R4567.
33. Monroe, C., Meekhof, D. M., King, B. E., Jefferts, S. R., Itano, W. M., Wineland, D. J., & Gould, P., (1995). Resolved-sideband Raman cooling of a bound atom to the 3D zero-point energy. Physical Review Letters, 75(22), 4011.
34. Monz, T., Nigg, D., Martinez, E. A., Brandl, M. F., Schindler, P., Rines, R., & Blatt, R., (2016). Realization of a scalable Shor algorithm. Science, 351(6277), 1068–1070.

35. Plašil, R., Zymak, I., Jusko, P., Mulin, D., Gerlich, D., & Glosík, J., (2012). Stabilization of H^+–H_2 collision complexes between 11 and 28K. Philosophical Transactions of the Royal Society A: Mathematical, Physical and Engineering Sciences, 370(1978), 5066–5073.
36. Raab, E. L., Prentiss, M., Cable, A., Chu, S., & Pritchard, D. E., (1987). Trapping of neutral sodium atoms with radiation pressure. Physical Review Letters, 59(23), 2631.
37. Ratschbacher, L., Sias, C., Carcagni, L., Silver, J. M., Zipkes, C., & Köhl, M., (2013). Decoherence of a single-ion qubit immersed in a spin-polarized atomic bath. Physical Review Letters, 110(16), 160402.
38. Schmid, S., Härter, A., & Denschlag, J. H., (2010). Dynamics of a cold trapped ion in a Bose-Einstein condensate. Physical Review Letters, 105(13), 133202.
39. Schmid, S., Härter, A., Frisch, A., Hoinka, S., & Denschlag, J. H., (2012). An apparatus for immersing trapped ions into an ultracold gas of neutral atoms. Review of Scientific Instruments, 83(5), 053108.
40. Sikorsky, T., Meir, Z., Ben-shlomi, R., Akerman, N., & Ozeri, R., (2017). Spin Controlled Atom-ion Inelastic Collisions. arXiv preprint arXiv:1709.00775.
41. Silvera, I. F., (1992). Prospects for Bose-Einstein condensation in atomic hydrogen and other gases. Journal of Low Temperature Physics, 89(1), 287–296.
42. Smith, W. W., Babenko, E., Côté, R., & Michels, H. H., (2003). On the collisional cooling of co-trapped atomic and molecular ions by ultracold atoms: Ca++ Na and Na 2+(v*, J*)+ Na. In: Coherence and Quantum Optics VIII (pp. 623–624).
43. Thompson, R. C., & Papadimitriou, J., (2000). Simple model for the laser cooling of an ion in a Penning trap. Journal of Physics B: Atomic, Molecular and Optical Physics, 33(17), 3393.
44. Webb, R. A., Washburn, S., Umbach, C. P., & Laibowitz, R. B., (1985). Observation of h e Aharonov-Bohm oscillations in normal-metal rings. Physical Review Letters, 54(25), 2696.
45. Weitenberg, C., Endres, M., Sherson, J. F., Cheneau, M., Schauß, P., Fukuhara, T., & Kuhr, S., (2011). Single-spin addressing in an atomic mott insulator. Nature, 471(7338), 319–324.
46. Wineland, D. J., & Itano, W. M., (1981). Spectroscopy of a single Mg+ ion. Physics Letters A, 82(2), 75–78.

47. Zhang, L., & Tanaka, H., (1998). Atomic-scale deformation in silicon monocrystals induced by two-body and three-body contact sliding. Tribology International, 31(8), 425–433.

Chapter 5

Introduction to Laser Cooling of Molecules

CONTENTS

5.1. INTRODUCTION

The movement of molecules within a room temperature gas is hundreds of meters per second. With these molecules, the precise measurements are hard to make, because their spectral lines are broadened with the broad range of Doppler shifts, and the observation time is limited by the high-speed limits. A vast range of vibrational and rotational states are also occupied by the molecules at room temperature, which tarnishes their quantum-mechanical characteristics (Shuman et al., 2010).

Back around 40 years, scientists have been tried to find means to cool *atoms* down to very low temperatures through laser light (Phillips, 1998). By the need for the improvement of spectroscopic measurements and atomic clocks, this beforehand work on laser cooling is motivated, but soon it has been realized that the applications expanded very much farther. Ever since the limelight of physics research has been the ultracold atoms. They are used to find out fundamental constants, test fundamental physics, investigate quantum phase transitions and quantum degenerate gases, study ultracold collisions, store, and process quantum information, simulate many-body quantum systems, and measure gravity, time, magnetic fields, acceleration, pressure, and electric fields (Anderegg et al., 2018).

Physicists are having much fun with atoms, and possibly the idea of molecules being an unhelpful complexity. Though, chemists are much more relaxed with molecules and are not convinced easily that it is interesting to investigate them at these low temperatures. Despite initially it seems that employing laser cooling techniques on molecules is extremely difficult. This case has been changed in last years. A method has been introduced by few groups to form ultracold molecules through binding ultracold atoms together (Moses et al., 2017). This method is proved to be highly successful for particular molecules, particularly those that are formed by alkali atoms, among gases of polar molecules within the quantum degenerate system now being manufactured this way (Marco et al., 2019). While the method to decelerate molecular beams at low speed is shown by the other groups, and then they have been confined those molecules for a lengthy time in storage rings and traps (Meerakker et al., 2012). The way to use laser cooling methods straight to the molecules is determined by a few groups, and now that has been completed with spectacular success (Shuman et al., 2009; Hummon et al., 2013). These days, the applications of such ultracold molecules are an insight to a diverse range of subjects. Some of such applications are discussed in the following passages.

To study fundamental physics, molecules are already being utilized in many ways (DeMille et al., 2017; Safronova et al., 2018). Heavy polar molecules are used by numerous groups to estimate the electron's electric dipole moment (edm). The Standard Model of particle physics anticipates this basic characteristic of the electron to be exceptionally tiny. Much greater values are predicted by the famous extensions of the Standard Model, e.g., supersymmetry, that are generally ten orders larger of magnitude. Even though they are quite tiny, these estimated values are between the approach of the very sensitive investigations. Molecules that are used in these experiments are polar in nature that improves the interaction energy within an applied electric field and the edm. The strict constraints are already made on theories by the recent measurements through warm molecules that broaden the Standard Model (Hudson et al., 2011; Baron et al., 2014). Likewise, exact measurement using molecules can investigate either the fundamental constants are varying with position or time (Hudson et al., 2006; Shelkovnikov et al., 2008). Theories that target to unify gravity with the alternate forces, and few theories of dark energy predict such a variation (Tokunaga et al., 2013). For the investigation of the effect of parity violation in chiral molecules and in nuclei molecules are being used as well. By applying ultracold molecules, the precision could be enhanced in all these studies to make better the degree of control (DOC) and widen observation times (Tarbutt et al., 2009; Tarbutt et al., 2013).

The improvement can also be brought in quantum science by ultracold molecules. Abundant electric dipole moments can belong to molecules that can be regulated easily, so they can make contact really strongly with one another using the long-range dipole-dipole interaction. By the accumulation of interacting quantum particles, notable new phenomena have emerged, e.g., superconductivity and magnetism. Such interacting many-body quantum systems are very complicated to simulate and are tough to inspect in solids in an organized way. Consequently, the phenomena and the means they turn up, are not fully understood. A perfect system to study the physics of quantum many-body systems are made by a regular array of polar molecules where each molecule is interacting with each other via dipole-dipole interaction, and with one molecule on every site of the array (Barnett et al., 2006; Micheli et al., 2006). For the formation of the array, the ultracold molecules can be laden into an optical lattice and an array of traps created by the interference of overlapping laser beams. Another area is quantum information processing, where molecules can have an influence, and many designs have been suggested (DeMille, 2002; Yelin et al., 2006).

5.2. BASICS OF LASER COOLING

Let's review the working of the Doppler cooling for atoms. Doppler cooling of a simple atom in a single dimension is illustrated in Figure 5.1(a). The atom has an excited state and a ground state, and ω_0 is the angular frequency of the transition within these two states. The transition possesses a matching natural linewidth $\Gamma = 1/\tau$, and the excited state gets a lifetime τ.

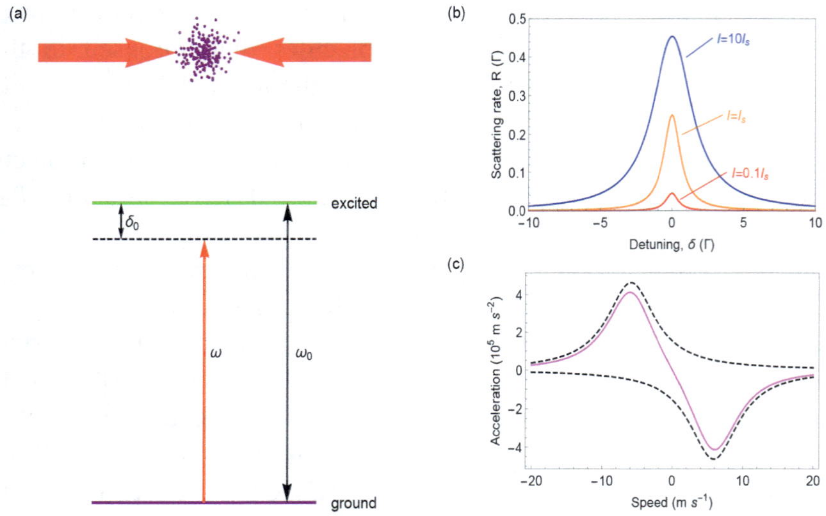

Figure 5.1. (a) Doppler cooling in 1D. A pair of similar counter-propagating laser beams have interacted with an atom. The laser frequency is somewhat lower than the atomic resonance frequency. (b) A rate of scattering as a function of detuning of laser light within the resonance, for three various intensities. (c) The acceleration of a sodium atom is a function of its speed. Dashed lines indicate the accelerations because of each of both beams, whereas the solid line signifies their sum. It is assumed that the beams possess intensities of I = Is, and are detuned by δ0 = −Γ by the λ = 589 nm transition (refers to the yellow line of sodium).

Source: https://www.tandfonline.com/doi/abs/10.1080/00107514.2018.157633 8?journalCode=tcph20.

It is observed that Γ is too small than ω_0, usually $\Gamma \sim 10^{-8}\omega_0$ – there is too sharp atomic resonances. The atoms make contact with a pair of similar counter-propagating laser beams having wavevectors $\pm k$. The laser angular frequency, ω, is tuned marginally less than the atomic resonance angular frequency; hence, it can be said that the light is 'red-detuned.' Detuning is the angular frequency difference, which is expressed as $\delta_0 = \omega - \omega_0$. Usually, detuning is selected to be of order Γ. It must be noted that this simplified model is normally enough, the model where the atom keeps only two energy levels (André et al., 2006). All the alternate levels of the genuine atom are not relevant as the light is blocked to resonance with a single specific transition, and very distant from resonance having all the others (Cairncross et al., 2017).

The two laser beams are not the same for an atom transmitting with velocity **v**, because of the Doppler shift $\delta_D = -\mathbf{k} \cdot \mathbf{v}$, which is opposite for both beams. With the absorption of a photon, a ground-state atom might be excited, probably from the laser beam that is moved opposite to the motion, as this light refers to Doppler-shifted nearer to resonance. The atom takes the momentum of such absorbed photon, lessening its velocity by ~*k*/*m*, where *m* indicates the atomic mass. Spontaneous emission makes the excited atom decay back to the ground state (Collopy et al., 2018). There is a random direction of the recoil of the atom connected to such photon emission. This series - preferred absorption out of the beam that resisted the motion pursued by randomly-directed spontaneous emission-iterates till *v* is close to 0. This procedure of absorption is referred, which keeps the linked force being the scattering force, and spontaneous emission being photon scattering (Barry et al., 2014; Williams et al., 2018).

The scattering rate for an atom contacting with a one laser beam having intensity *I* is:

$$R = \frac{\Gamma}{2} \frac{I/I_s}{(1 + I/I_s + 4\delta^2/\Gamma^2)}, \tag{1}$$

where $\delta = \delta_0 + \delta_D$ indicates total detuning, containing the Doppler shift. The quantity I_s is called saturation intensity and is expressed as $I_s = \pi hc\Gamma/(3\lambda^3)$ where λ signifies the wavelength. $I \ll I_s$, R grows linearly along with *I* at short intensity, on the other hand, $I \gg I_s$, R saturates about the largest value Γ/2 at immense intensity. Eq. (1) is plotted in Figure 5.1(b) being a function of δ for some of the various values of I/I_s. Commonly, for δ = 0, the scattering rate is maximum, then corresponding to atom, the light is resonant. The analysis is a Lorentzian having a total width at the half utmost of Γ Simply, this is the natural linewidth concerning atomic transition at low intensity, whereas it is broadened at high intensity as rather than the wings, the scattering rate is more powerfully saturated at the mid of the line. The scattering force can be worked out by this reliance of R on δ, and the reason for the dependence of it on the speed of the atom. Force is defined as the rate of change of momentum, and a momentum ~k is transferred by every scattering event, therefore, the scattering force because of a single beam is F = ~kR. For two counter-propagating beams each having an intensity I, there are opposite forces from every beam but do not completely cancel because of the opposite Doppler shifts. On $I \ll I_s$, the total force is the total (sum) of both individual forces. The related acceleration is shown in Figure 5.1(c) as a function of speed, for the instance of sodium atoms. Laser detuning of $\delta_0 = -\Gamma$ can be chosen in each beam along with intensity ($I = I_s$). The accelerations because of the two individual beams are shown by two dashed lines, and the solid line indicates the total (sum) of these. It is seen that the accelerations are maximum, up to 4×10^5 m s^{-2}. For slow speeds, there is a linear change of acceleration with speed and possesses the opposite sign to the speed, which means that the motion is opposed (Zhelyazkova et al., 2014; Cheuk et al., 2018).

At small speed, a Taylor expansion of $F(v)$ can be made around $v = 0$, and hence, found $F \approx -\alpha v$, where:

$$\alpha = \frac{8\hbar k^2 \delta_0 I/I_s}{\Gamma(1 + I/I_s + 4\delta_0^2/\Gamma^2)^2}, \qquad (2)$$

Signifies the damping constant. This force damps the atomic velocity towards zero. There is a too high acceleration that the damping becomes really rapid (in case of sodium), there are just 14 μs of the characteristic damping time. Counter-propagating beams with each of the three orthogonal axes are introduced to expand the cooling into ternary dimensions. Therefore, all three elements of the velocity are absorbed toward zero. This occurs a three-dimensional optical molasses.

The spontaneous emission takes place in a random direction in every scattering event. When there is almost zero speed, absorption turns equally likely from all of the beams, thus, the absorption events hold random directions as well. Hence, two momentum kicks of ~k are got by the atom for each scattering event, and each of them is randomly directed. This random scrambling of the atom is known as a heating mechanism. It counteracts the damping in the direction of zero velocity. There exists a Doppler temperature, which is an equilibrium temperature, where the cooling and heating are balanced (Gorshkov et al., 2011; Blackmore et al., 2018). The expression of the Doppler temperature is:

$$T_{\rm D} = -\frac{\hbar\Gamma^2}{8k_{\rm B}\delta_0}\left(1+\frac{I}{I_{\rm s}}+\frac{4\delta_0^2}{\Gamma^2}\right). \quad (3)$$

When $\delta_0 = -\Gamma/2$ and $I \ll I_s$, the lowest temperature is obtained, $T_{D, min} = \sim\Gamma/(2k_B)$, called the Doppler limit. Its value for sodium is 240 μK. It will be discussed in section 3.5, there are alternate cooling mechanisms that can take temperatures less than this limit, and atoms are usually laser-cooled to fewer temperatures with these sub-Doppler methods.

5.3. APPLYING LASER COOLING TO MOLECULES

Upon the absorption of a photon, the change in velocity is called the recoil velocity, which is expressed as $v_r = \sim k/m$. The value for sodium is $v_r = 0.03$ m/s, and it is similar to other molecules and atoms. ~300 m/s is the usual speed for molecules or atoms in a room-temperature vapor, so a minimum of 10^4 scattering events are taken to lead this speed near to zero. A "closed" transition is required for that, which means that the cooling cycle cannot be left by the excited atom through decaying to any intermediate state which is not supposed to resonant with the laser light. Some atoms take these closed transitions; thus, they are smooth to cool. Molecules have the supplementary complication of rotational and vibrational energy levels. This energy level structure is needed to be examined to know how the method to apply laser cooling to molecules (Yan et al., 2013).

5.3.1. Energy Level Structure

The energy level structure of a conventional diatomic molecule is illustrated in Figure 5.2(a). Similar to atoms, molecules also possess a set of electronic states. The labels of the ground electronic state are always *X*, whereas, the labels of the excited electronic states are alphabetical, *A*, *B*, *C*, etc, arranged by energy. The electronic energy, E_e, for every electronic state is a function of the distance within the nuclei, *R*, therefore, each electronic state is mapped as a curve $E_e(R)$. This is called the potential energy curve, as it facilitates being the potential where the nuclei move. The nuclei move back and forth about the equilibrium separation, R_0, turning to a set of quantized vibrational energy levels, marked by the quantum number, i.e., $v = 0,1,2$. With the solution of the one-dimensional Schr¨odinger equation for the sake of the potential energy curve $E_e(R)$, these vibrational energy eigenvalues, E_v, and the related eigenfunctions, $\Phi_{e,v}(R)$, are obtained. Close to the lowest potential energy curve, the motion is near to the one of a harmonic oscillator, hence, the vibrational wavefunctions are the matching harmonic oscillator eigenfunctions, and the vibrational energy levels comes about $E_v \approx (v+1/2)\sim\omega_{vib}$. Some instance vibrational wavefunctions are shown in Figure 5.2(b), whose detailed discussion is as under. There is a set of rotational states for every vibrational state (Andreev & Hutzler, 2018). Think of from the classical mechanics, that $L^2/2I$ is the energy of a rotating body, where I represent the moment of inertia of the body and L indicates the angular momentum. The continuous variable L^2 is replaced with $\sim^2 N(N + 1)$ by quantization of the angular momentum, where $N = 0,1,2.$ represents the rotational quantum number, hence, the rotational energies are expressed as $E_r = BN(N + 1)$ where $B = \sim^2/2I$ is called the rotational constant of the molecule. As compared to the spacing of vibrational levels, typically, there is a much smaller spacing of rotational levels, which consequently is much lower as compared to the spacing of the potential energy curves analogous to various electronic states. This hierarchy is summarized as $\Delta E_r \ll \Delta E_v \ll \Delta E_e.$

5.3.2. Vibrational Branches

If the energy level of the structure is known, the way to use laser cooling to molecules can be considered. Consider a transition driving from the electronic ground state, i.e., X, having rotational and vibrational quantum numbers N and v, marked |gi = |X, v, Ni, to a particular rotational (N') and vibrational (v') level of the initial excited electronic state, A, marked |ei = |A, $v'N'$i. For efficient laser cooling, |ei should decay particularly to |gi therefore, the molecule continues to be resonant to the laser light, enabling it to disperse a large number of photons from the laser. Unluckily, that does not happen normally. Rather, |ei can decay to either of the various vibrational states of X. For decaying to v" the probability is:

$$P_{v',v''} = \left| \int \Phi_{X,v''}(R)\Phi_{A,v'}(R)dR \right|^2 . \tag{4}$$

It is the square of the overlap integral within two vibrational wavefunctions, one from state A, and one from state X, and is called the Franck-Condon factor (Altuntaş et al., 2018).

For the state X potential, the starting five even-numbered vibrational wavefunctions, i.e., $\Phi_{X, v}$" that is demonstrated in Figure 5.2(a) are illustrated in Figure 5.2(b) (in blue). The smallest vibrational wave function of state A, $\Phi_{A, v'=0}$, is also shown for each one (in green).

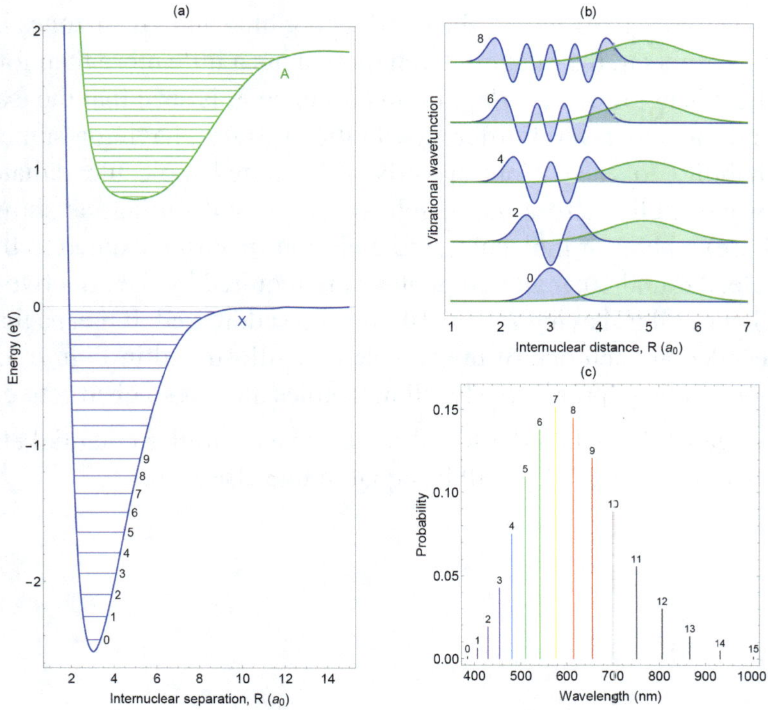

Figure 5.2. (a) Potential energy curves of the ground electronic state (X) and the earliest electronically-excited state (A) of a conventional diatomic molecule - for instance, LiH. The vibrational energy levels for each potential are illustrated. (b) Minimum vibrational wavefunction (v' = 0) for the state A (indicated by green), and a selection (v" = 0,2,4,6,8) of vibrational wavefunctions for the state X (indicated by blue). The Franck-Condon factor is given by the square of the overlap integral within a vibrational wavefunction corresponding to the X state and one from the A state. (c) Emission spectrum with respect to the molecules excited to the v' = 0 vibrational level of the state A.

Source: https://www.tandfonline.com/doi/abs/10.1080/00107514.2018.1576338?journalCode=tcph20.

A molecule that is excited to this state is decayed to different vibrational states of X having probabilities $P_{0,\ v''}$ that is given by equation (4). The overlapping blue and green areas in the last row of Figure 5.2(b) show the overlap integral for $v'' = 0$. As there is a different equilibrium bond length of both potential curves, so there is not a great overlap, however, just the wings of both ground-state wavefunctions are overlapped. Consequently, there is a small $P_{0,0}$. For $v''= 2$, the overlapping blue and green zones in the later row up refers to the integral. Even though it's a little more than for $v''= 0$, still, there is not much overlap. In this instance, it is seen that the excited molecule decays to several different vibrational states, having a somewhat less probability to each. Consequently, the excited molecule radiates at various wavelengths across the visible spectrum and further, as shown in Figure 5.2(c) where for molecules, the emission spectrum excited to the $|A, v'= 0$i state. Around 10^4 scattered photons are required by the effective laser cooling, hence all v'' having $P_{0,v''} > 10^{-4}$ must be addressed. It means that the laser is needed at each one of the wavelengths illustrated in Figure 5.2(c). This is not practical, prosperously, all molecules are not so obstructive.

It is obvious that when the A and X states have similar potential energy curves, then, if $v''= v^0$, $P_{v',v''}$ will be equal to one else zero.

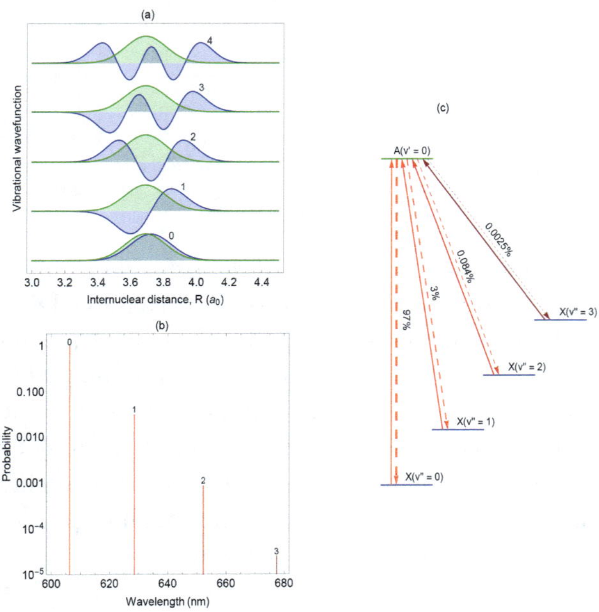

Figure 5.3. Some molecules, like CaF, possess vibrational branching ratios appropriate to laser cooling. (a) A selection (v" = 0,1,2,3,4) of vibrational wavefunctions corresponding to state X (refers to blue), and lowest vibrational wavefunction (v' = 0) corresponding to state A of CaF (refers to green). (b) Emission spectrum for molecules excited to the v' = 0 vibrational level of state A. Observes the logarithmic scale. (c) Laser cooling design for CaF covering four lasers.

Source: https://arxiv.org/abs/1902.05628.

In this situation, a vibrational state of *A* can only decay to a single vibrational state of *X*. Some molecules closely meet this ideal. Usually, they are ones where the molecular bond does not involve that electron that gets excited in the transition. Then, a very little change is involved by the transition to either the strength or length of the bond, and there will be similar potential energy curves for the excited and ground states. A good instance of such a molecule is Calcium monofluoride, CaF. To a good estimation, for the formation of the ionic bond, one of the two valence electrons of calcium are looked at the fluorine, whereas the other stand localized on the calcium, and is the electron that becomes excited in the electronic transition within the *A* and *X* states. In the X state of CaF, the wavefunctions for the smallest five vibrational levels ($v'' = 0$ to 4) are shown in Figure 5.3(a) (in blue), while the ground vibrational wavefunction ($v^0 = 0$) of the state *A* is shown (in green). As there is a similar potential energy curve of *A* and *X*, hence, there is a big overlap between the $v'' = 0$ and $v' = 0$ states. However, there is less overlap with all other states. For molecules excited to |*A*, $v' = 0$i, the resulting emission spectrum is shown in Figure 5.4(b). Observe the logarithmic scale at the vertical axis, which is spanning five orders of magnitude.

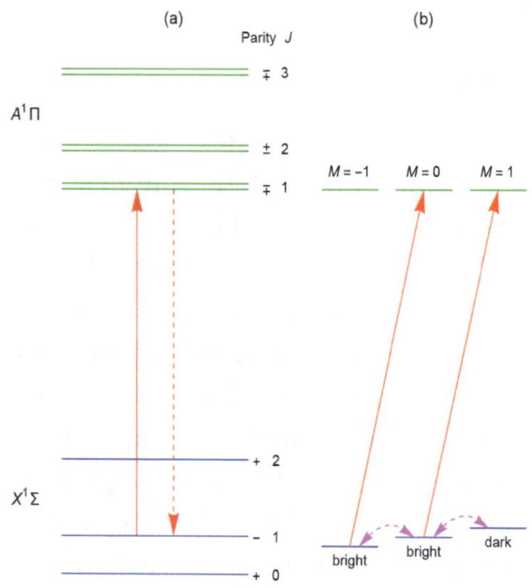

Figure 5.4. (a) For a diatomic molecule, the rotational structure of a conventional laser cooling transition (for example, any one of the transitions shown in Figure 5.3(c)). The labels indicate the total angular momentum (J) and parity for each state.

Source: https://arxiv.org/abs/1902.05628.

Note: The transition determined by the solid arrow refers to "rotationally closed." (b) Zeeman sub-levels corresponding to the J = 1 upper and lower states. The element of angular momentum with the z-axis is stated by M. The transitions are shown by the arrows that can be directed by left-handed circularly-polarized light transmitting with the z-axis. For this selection of polarization, the M = 1 state is considered to be a dark state. A different dark state will be produced by a different polarization choice. The Zeeman sub-levels in a magnetic field have diverse energies, and there is an unstable dark state which is a superposition of these states (shown by magenta arrows).

It is seen that $P_{0,0} \approx 1$, and also $P_{0,0} \gg P_{0,1} \gg P_{0,2} \gg P_{0,3}$. This is one of the important ingredients using for practical laser cooling of molecules (Rosa, 2004). In Figure 5.3(c), an instance of CaF is shown in which four lasers are applied to focus the transitions from $v'' = 0, 1, 2, 3$. Then, the leftover leak out of the cooling cycle is so small that molecules will scatter, on average around 2×10^5 photons, enough for powerful laser cooling.

The problem of decay to many vibrational states is solved by the usage of many lasers to sermon sufficient vibrational transitions, and a cautious selection of molecule.

5.3.3. Rotational Branches

There exists a ladder of rotational states inside each vibrational state. Luckily, the parity selection rules and the angular momentum limit the amount of accessible rotational states. The excited states associated with the laser cooling transition, and the conventional rotational structure of the ground are shown in Figure 5.4(a). A case is considered to keep things comparatively simple, in which there is no unpaired nuclear spin and no unpaired electron spin. Next, the angular momentum occurs only from the rotation of the molecule as a whole and from the orbiting electrons. Usually, the projection of the electronic orbital angular momentum upon the internuclear axis is referred due to the molecule's symmetry (Weidemüller et al., 1994). In the example given here, and in the units of ~, this is 1 for the *A* state and 0 for the *X* state. These are called Π and Σ states and are similar to the P and S states of atoms. To produce the total angular momentum, the rotational and orbital angular momenta are added together. As the rotational ladder is gone up, thus, in the *X* state, the parity changes within negative and positive, and the minimum level has $J = 0$. The minimum level in the *A* state has $J = 1$ because this is the lowest value of J that can contribute to a projection of 1 upon the internuclear axis. There are a pair of states having opposite parity, i.e., *p*, for every value of *J* in state *A*. That's why, the projection of orbital angular momentum upon the internuclear axis can be either -1 or +1, and the energy eigenstates are the antisymmetric and symmetric linear combinations of such possibilities.

Let the parity and angular momentum be pointed out using the notation J^p. For electric dipole transitions, the requirement of the selection rules is that the parity alters in the transition and that $\Delta J = 0,\pm 1$. Initially, consider the only permitted transition coming from the smallest level of *X*. 1^- is the excited state of such transition. This state can deteriorate back to 0^+, however, can decay to 2^+ as well. The existence of both decay channels is not convenient for laser cooling, therefore, it makes twice the number of needed laser frequencies. Rather, consider the transition that is from 1^- to 1^+. Hither, the above state can decay back just to the 1^- level of *X*, hence, it is a "rotationally closed" transition, which means that a single rotational element requires to be addressed (i.e., for every vibrational state) (Stuhl et al., 2008). These rotationally closed transitions have been made use by all laser cooling work to date.

5.3.4. Dark States

One last problem that is needed to be solved. In Figure 5.4(a), the laser cooling transition is indicated having $J = 1$ in both the excited and the ground state. Each possesses three Zeeman sublevels, having quantum numbers $M = 0, -1, +1$, as illustrated in Figure 5.4(b). The angular momentum of the atom is expressed by the *M* quantum number in the *z*-direction. Assume a circular polarized laser beam that is propagating with the *z*-axis. There is a single unit of angular momentum by the photons of this beam along the *z*-axis. With the absorption of a photon, if the atom is excited, *M* should increase by 1 to save the angular momentum along the *z*-axis, but it is impossible for molecules having $M = +1$, as the excited state does not possess an $M = +2$ state. Any of the photons cannot be scattered by the molecules in $M = +1$ from the laser, but once they approach this state they stay there endlessly. It is called a dark state. The laser can excite the molecules in other states, however, the excited state can be decayed to $M = +1$. Hence, each molecule will stop at the dark state subsequent to scattering only a few photons, avoiding laser cooling. This problem is illustrated for a single choice of polarization. If the circular polarization is chosen with the diverse handedness, then $M = -1$ state would become dark, and if the light linearly polarized along the *z*-axis is chosen, then $M = 0$ state becomes dark. Generally, for any selection of laser polarization, there will be a rational superposition of the ground state sub-levels, i.e., a dark state, and always molecules will approach this dark state after dispersing only a few photons (Dalibard & Cohen, 1989).

There are two possible solutions to this problem. One is to change the polarization of light within two orthogonal states, for instance, from left-circular to right-circular. If the state is dark corresponding to one polarization, then it will not be dark corresponding to the other, hence, if the modulation is achieved sufficiently rapidly, then the molecules will never lie in a dark state protractedly. The related timescale is the time consumed to approach the dark state, which is around 1 μs in a typical investigation. A second solution is to select the polarization, therefore, the dark state becomes a superposition state, and use a short magnetic field in order to 'de-stabilize' that superposition. The energy relies on M in a magnetic field, thus, the states that set up the superposition have diverged energies (refers to Figure 5.4(b)). In correspondence to the time-dependent Schrodinger equation, the state $|1i+e^{-i\Delta t/\sim}|2i$ is evolved by a starting dark superposition state $|1i + |2i$ after a time t, and Δ refers to the energy difference within states $|1i$ and $|2i$. The new state does not become dark any longer due to such a change of phase, hence, the molecules will not lie in a dark state for a long period given Δ is sufficiently large. In many cases, a slight magnetic field, i.e., sometimes the Earth's field, is sufficient to destabilize efficiently either dark state.

5.3.5. Doppler and Sub-Doppler Cooling

The method for laser cooling is clear now. Select a molecule having strong electronic transition and that has convenient vibrational branching ratios, apply adequate vibrational repump lasers to minimize the leak out of the cooling cycle to an agreeable level (normally around 10^{-5}), propel the rotationally closed transition shown in Figure 5.4, and by using a magnetic field or switching the polarization of the light, de-stabilize the dark states. Then, the contact of the molecule with the light is really alike to that of the simple two-level system pictured in Figure 5.1. Specifically, the formulae for the damping constant, scattering rate, and Doppler temperature are roughly unchanged. It can be expected for red-detuned light that Doppler cooling can cool the molecules near to the Doppler limit, similar to atoms.

Practically, lessen temperatures, as compared to the Doppler limit, are usually reached for atoms, that is because there are alternate cooling methods, different from Doppler cooling, that become useful for low temperatures. These are called sub-Doppler cooling approaches, and there are numerous different kinds (Dalibard & Cohen, 1989; Ungar et al., 1989). In the most useful of them, dark states are knowingly exploited and set up, indicating that despite the chance of problematic dark states for laser cooling, they can, in addition, be put to valuable use (Weidemüller et al., 1994). Figure 5.5 illustrates the technique, where for simplicity, a molecule is considered with two ground sub-levels and a single excited state. A molecule can be excited by the laser light from one of these sub-levels (i.e., the bright state), though not from the other (i.e., the dark state). Now assume a pair of counter-propagating laser beams, that must be marginally detuned from resonance. If there is the same polarization for the two beams, they arranged an intensity standing wave having a period of $\lambda/2$. For the orthogonal polarizations, there is a uniform intensity, but the local polarization alters with position, having a period of λ. When the polarization of the two beams are neither perpendicular nor parallel, both the local polarization and the intensity transform with the position. Light influences the energy levels of the molecule. The electrons within the molecule are driven by the oscillating electric field of the light, making an electric dipole that is oscillating at the same frequency. A minimal shift of the ground-state energy levels is occurred due to the interaction within the electric field of the light and this dipole, positive on the blue-detuned light and negative on the red-detuned light. This is called the ac Stark shift or the light shift. This influence shifts the energy of the bright state, however, the dark state does not shift as it does not make contact with the light. Figure 5.5 illustrates the case of blue detuned light, hence the bright state is situated above the dark state. The shift of the bright state alters with the position in the standing wave, being the largest on the greatest intensity. Assume a molecule within the bright state, close to a node of the standing wave. Its internal energy rises because of the ac Stark shift as it moves in the direction of the antinode, hence, its kinetic energy should be decreased. The slowing down of the molecule is considered as it mounts the potential hill. The molecule mostly seems to be excited at the time when it is closed to the antinode, hither, the light is extremely intense. Thus, it turns to be optically pumped in response to the dark state when it is closed to the top of the potential hill. Two processes can transfer back the molecule to the bright state closely associated with both methods for de-stabilizing dark states mentioned in the given below sections (Shuman et al., 2009). The dark state will be changed into a bright state when there is an appropriate size magnetic field. Rather the molecule can be transferred back to the bright state by the movement of the molecule from the altering polarization of the light. When the dark and bright states are close in energy, which is close to the bottom of the potential hill, then this change is most likely to occur. Consequently, the molecule turns to follow the purple arrows shown in the figure, dropping energy because it mounts the potential hills repeatedly. This is referred to as Sisyphus cooling. Moreover, as the molecules employ a lot of their time within the dark state, there is a reduction of the photon scattering rate, and also of the heating because of the arbitrary jostling of photon scattering events. Hence, this is an effective approach for cooling temperatures less than the Doppler limit.

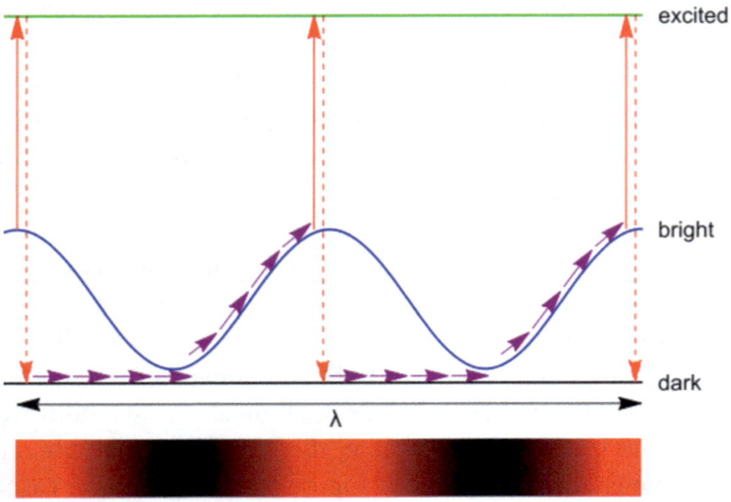

Figure 5.5. The Sisyphus cooling scheme.

Source: https://en.wikipedia.org/wiki/Sisyphus_cooling.

5.3.6. Progress in Laser Cooling of Molecules

A team at Yale has laid the groundwork on laser cooling of molecules with strontium monofluoride (SrF) molecules. Corresponding to this molecule, there is 98% branching ratio of the main transition, $A(v^0 = 0) \leftrightarrow X(v^{00} = 0)$. Using a single vibrational repump laser sealing the leak to $v^{00} = 1$, it is shown by the team that the molecules are scattering around 150 photons within a 50 μs interaction time, adequate to alter their velocity to 0.8 m/s (Shuman et al., 2009).

After that, a second repump laser is added to seal the leak to $v^{00} = 2$, therefore, many more photons could be dispersed. Applying this laser light, a one-dimensional optical molasses is formed, and it is used to cool a beam of SrF molecules to one transverse direction (that is one of the directions which is perpendicular to the propagation direction of the beam) (Shuman et al., 2010). Figure 5.6(a) illustrates the experiment. For the synthesis of a molecular beam, a cryogenic buffer gas source has been utilized, and the laser beams have been reflected back and forth at a minimal angle, therefore, they have intersected the molecular beam a lot of times, making a 15 cm lengthy interaction region. Subsequent to mentioned cooling region, the molecular beam's density distribution is measured. The outcomes of these experiments are shown in Figure 5.6(b) for different cases. A 0.5 mT magnetic field used to destabilize dark states, and the outcome for a red-detuned cooling laser ($\delta_0 = -1.5\Gamma$) are shown in the bottom left plot. The distribution is narrower for the on laser light (refers to red curve) as compared to the off one (refers to the black curve), as Doppler cooling takes molecules to lower transverse speeds, minimizing the diversity of the beam. A temperature is assigned by the scientists that are a property of the transverse velocity spread, and predict that the Doppler cooling has minimized this temperature, i.e, from 50 mK to around 5 mK. The top left plot illustrates the happening that is done on the blue-detuned cooling laser ($\delta_0 = 1.5\Gamma$). In this case, the divergence of the beam is increased due to the Doppler heating, hence, its size is increased at the detector. The right plot demonstrates the change of distributions on the reduction of the magnetic field to 0.06 mT. In this scenario, when the light is blue detuned, and at low transverse speeds, there is powerful inflation of molecules, which is the property of magnetically-assisted Sisyphus cooling. It is estimated by scientists that the transverse temperature is minimized to around 300 μK. There are two peaks for red-detuning that are symmetric nearby the center, which conform to molecules moved towards an equilibrium non-zero velocity. Because of competition between Sisyphus heating and Doppler cooling, it occurs.

It is shown from the phenomenal results that for molecules, laser cooling could work actually. After that, the same method has been used by some other groups to other molecules. Transverse Doppler cooling of a beam of YO molecules is expressed in one and two dimensions, decreasing the temperature to around 15 mK (Hummon et al., 2013). A beam of SrOH is cooled to 750 μK in a single transverse direction applying the Sisyphus mechanism, which shows that laser cooling is also applicable to polyatomic molecules (Kozyryev et al., 2017). A beam of YbF molecules is cooled in a single dimension using similar methods to a temperature less than 100 μK (Lim et al., 2018). This molecule is being applied for the measurement of the electron's electric dipole moment (Hudson et al., 2011). The capacity to cool it to this low temperature has great prospects for enhancing the precision of this measurement (Tarbutt et al., 2013).

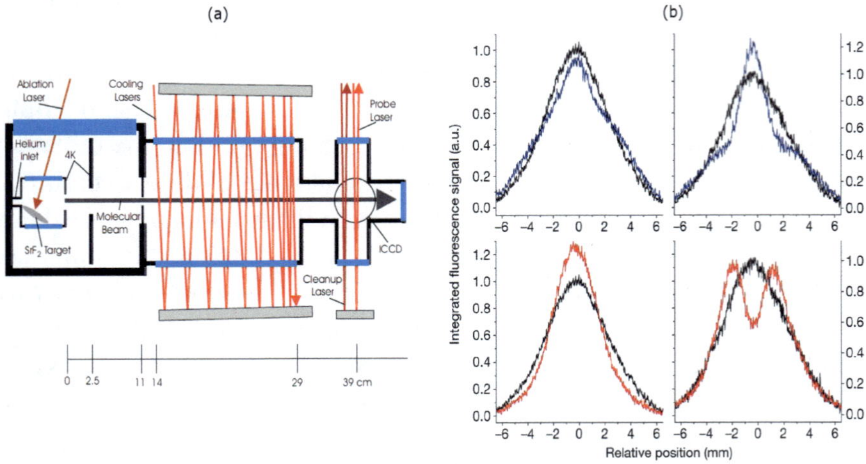

Figure 5.6. Transverse laser cooling concerning a SrF molecular beam. (a) Experimental setup. A molecular beam, originating from a cryogenic buffer gas, traversing to one-dimensional optical molasses produced by numerous passes of the laser light, and is then portrayed onto a camera (b) Consequently density distributions of the molecular beam.

Source: https://www.nature.com/articles/nature09443.

Note: Red curves: detuning of $\delta_0 = -1.5\Gamma$; black curves: absence of cooling light; blue curves: detuning of $\delta_0 = +1.5\Gamma$. Right: for 0.06 mT magnetic field, magnetically-assisted Sisyphus cooling is noticed for blue-detuning (upper plot), and Sisyphus heating is observed for red-detuning (lower plot). Left: for 0.5 mT magnetic field, Doppler cooling is noticed for red-detuned light (lower plot) and Doppler heating is observed for blue-detuned light (upper plot). With permission from Springer Nature, adapted from reference (Shuman et al., 2010).

Molecular beams have been used by these transverse laser cooling experiments, where there are just a few m/s of the initial spread of the transverse velocities. It is shown by the experiments that the low temperatures could strike, but it only changes the velocities by this trivial amount. The molecules are still dispersing at high speed in a forward direction, generally 100–200 m/s, along with velocity spreads of about 30 m/s. Another major challenge has to apply the radiation pressure of the laser light to slow down these molecular beams until rest. Minimum 10^4 photons are required generally to be dispersed from a counter-propagating laser beam. Only those molecules that remain within the laser beam will be decelerated, however, the molecular beam is straying, with increasingly molecules leaving the area of the laser light with the passage of time. Hence, the possibly fast deceleration of the molecular beam is significant, which demands a great scattering rate. This calls for rapid de-stabilization of dark states, high laser power, and ideally tuned laser frequencies. The molecules are moving towards the laser, therefore, look at a Doppler shift in the direction of a higher frequency. For a typical speed of 150 m/s, and 600 nm wavelength, the Doppler shift is about 250 MHz, which is approximately 40 times greater than the conventional linewidth of the laser cooling transition. As the laser frequencies can be adjusted for compensation, so this shift itself is not an issue. But the Doppler shift alters with the deceleration of the molecules. Usually, a velocity change of about 5 m/s alters the Doppler shift by approximately a natural linewidth ($\delta_D \approx \Gamma$). Referring to equation (1), or Figure 5.1(b), and considering $I = I_s$ as an instance, it is observed that this detuning ($\delta = \Gamma$) minimizes the dispersion by a factor of 3. After slowing down a little, the molecules are observed no longer resonant with light, hence, stop dispersing photons and stop deceleration (Kozyryev et al., 2017).

One solution to this problem is to widen the frequency spectrum of the laser by itself. The lasers applied for laser cooling are remarkably monochromatic. Usually, the light has a frequency spread less than 1 MHz, which means that the frequency is expressed with a precision improved than 1 part per billion. Normally, this is desirable because it set all the power upon the frequency that signifies. Nevertheless, in order to decelerate the molecules, it assists to widen the frequency spread of the laser to around 250 MHz, therefore, the molecules can decelerate until rest without dropping to resonance. Sometimes, this method is known as frequency-broadened slowing, and sometimes it is called white light slowing. The team at Yale, after the demonstration of the first laser cooling of molecules, has decided to slow down their SrF beam with the help of the white light slowing technique. An electro-optic modulator has been used to adjust the frequency of the laser, which adequately broadens its spectrum. Beginning from a beam having 140 m/s mean speed, about 6% of the molecules having speeds less than 50 m/s can slow (Barry et al., 2012).

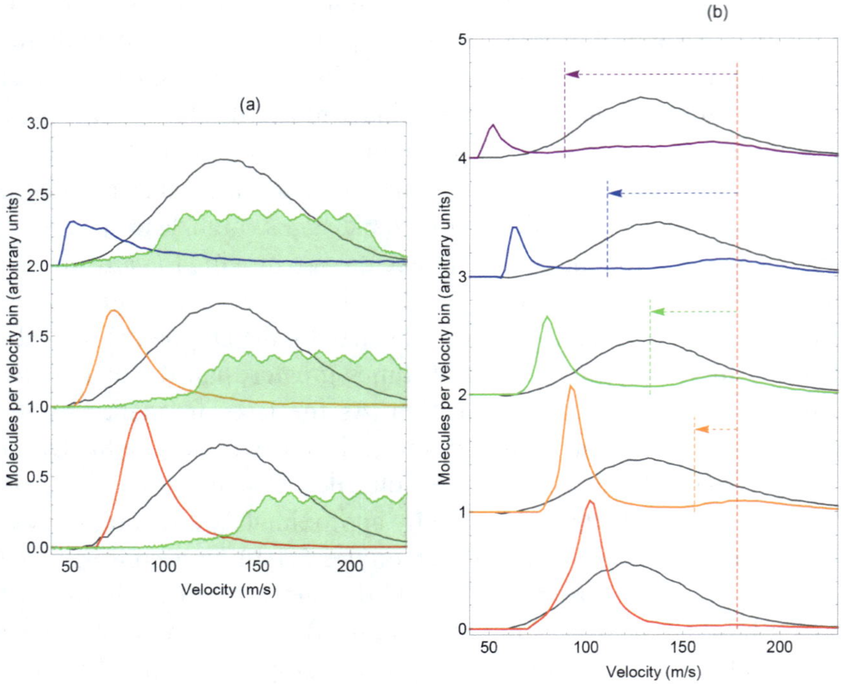

Figure 5.7. Decelerating molecular beams by radiation pressure. Velocity distributions of CaF molecules without (indicated by gray lines) and with (indicated by colored lines) the slowing light used are shown in the plots. Two various slowing methods are illustrated. (a) Slowing with a 6 ms pulse of frequency-broadened light. The frequency spectrum of the light switched onto the velocity scale is shown by the green shaded area. (b) Slowing with a 6 ms pulse of frequency-chirped light. The frequency of the light at the beginning and end of the chirp, which is transposed onto the velocity scale is shown by the vertical dashed lines.

Source: https://iopscience.iop.org/article/10.1088/1367–2630/aa5ca2/meta.

The frequency can be altered with the passage of time despite the widening the frequency spectrum of the light, in order to make up for the fluctuating Doppler shift since the molecules are slowed down. This method which is also known as frequency-chirped slowing has been used at Imperial College London for the deceleration of beams of CaF molecules (Zhelyazkova et al., 2014; Truppe et al., 2017). There has also been also a comparison of the frequency-chirped method and the frequency-broadened method. The outcomes of such experiments are shown in Figure 5.7. where the velocity distributions of the CaF beam is shown along with and without the application of slowing light. The frequency-broadened method is applied in Figure 5.7(a). Around 220 MHz, the spectrum of the laser light is widened, with respect to a velocity extent of around 120 m/s. Three datasets are illustrated. In the lowest one, there is a resonant central frequency of the light, along with molecules exciting at around 200 m/s. The molecules are decelerated and are gathered at about 90 m/s. In the middle and upper graphs, the light is switched to lower frequencies and causes the deceleration of molecules to lower velocities. A series of experiments are shown in Figure 5.7(b) that applies the frequency-chirped method. In the last plot, no chirp has appeared, and there is a resonant fixed frequency of the light along with molecules exciting at 178 m/s. They are slowed down by the light and gather at about 100 m/s. The next plots indicate the happening when there is a swept linear frequency all along the 6 ms period though the light is on. In each case, the span of the sweep is demonstrated by the arrows. The molecules are made to move to lower velocities because the light is passed to lower frequencies. There are narrow decelerated velocity distributions, which shows that the molecules are cooled and also slowed. Due to the deceleration of molecules to the lower speed, the number of molecules is decreased, as the transverse velocity spread is not changed, hence, the fewer molecules approach the area of the detector and the slower beams diverge too. This climax the importance of the rapid deceleration of the beam. Subsequent to the optimization of the frequency-chirped method, the group at Imperial has been able to slow down around 10^6 CaF molecules to 15 m/s. Similar methods have been used at Harvard for CaF molecules and at JILA for YO molecules (Yeo et al., 2015; Hemmerling et al., 2016). Recently, that laser slowing to low speed and laser cooling to low temperature has been shown, and the attention is diverted to the issue of trapping the molecules.

5.4. MAGNETO-OPTICAL TRAPPING OF MOLECULES

A velocity-dependent force is provided by optical molasses, which reduces the velocity of the atoms towards zero. This does not cause a trap on its own, and the molecules or atoms will diffuse away slowly. To produce a magneto-optical trap (MOT), a magnetic field gradient is added to the optical molasses which offers a position-dependent force. It is described below with more details. By this, atoms are trapped for long periods and are cooled to lower temperatures. For the past 30 years, the MOT has confirmed spectacularly beneficial. It has been the initial point for nearly all the innumerable researches using ultracold atoms. The same is for the ultracold molecules.

5.4.1. Principles of the Magneto-Optical Trap

The principle of MOT is illustrated in Figure 5.8. The pairs of circularly-polarized, red-detuned laser beams are counter-propagated with each of the coordinate axes and producing optical molasses. A magnetic field is created by equal but opposite currents moving in a pair of coils. The arrows within the region indicate the direction and strength of this field where the beams overlap. At the center, the field is zero which grows linearly in all directions off the center, points off the center in the horizontal direction, and points close to the center in the vertical direction. The long red arrows indicate the propagation directions of the beams, and the short magenta arrows indicate the directions of the photon spin in each beam (Zhelyazkova et al., 2014; Truppe et al., 2017).

Assume an atom at someplace in the region where there is an intersection of the beams. Here, it is supposed that the atom possesses a ground state having angular momentum $F = 1$ and an excited state having $F = 2$ and that the excited state holds a magnetic moment whereas the ground state does not. Let's assume the atom is moved from the center, towards horizontally to the right. Here, the magnetic field is pointed towards the right, and the z-axis direction can be chosen. The magnetic field and the Zeeman splitting of the excited state increase with the increase in the z-axis. A photon arising from the right involves its spin along $-z$, hence, absorption of this photon must decrease M by 1 to conserve angular momentum. On the other side, a photon arising from the left contains its spin along $+z$, so M makes rise by 1 on the absorption of this photon. Due to the negative detuning of the light and the Zeeman shift, transitions that decrease M are nearer to resonance, thus, the atom is supposed to disperse photons from the right. The atoms are pushed back towards the center due to the presence of net radiation pressure. However, if the atom is displaced towards one of the other beam directions, then again the z-axis is chosen in the line of the local magnetic field, and it is seen that the description is not changed. The photons whose spin is adverse to the direction of the magnetic field, are absorbed preferentially by the atoms due to the Zeeman splitting, and these photos are always originated from the beam that forces the atom towards the center, the place where the field is zero.

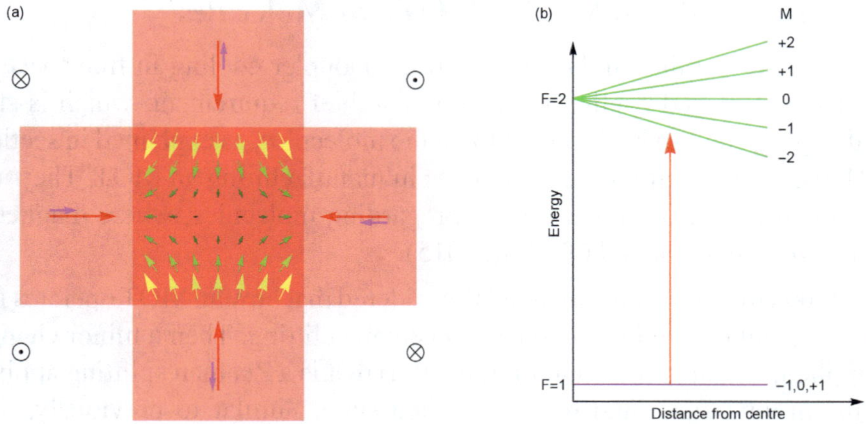

Figure 5.8. Magneto-optical trapping. (a) Two counter-propagating, circularly-polarized red-detuned beams pass along all three coordinate axes. (b) Energy levels of a model atom being a function of displacement, z, with any of the k-vectors.

Source: https://ui.adsabs.harvard.edu/abs/2019arXiv190205628T.

Note: (a) Pink region shows the two pairs. Their k-vectors are shown by long red arrows, and for each beam, the directions of the photon-spin are indicated by short magenta arrows. A pair of coils carry equal but opposite currents upon their axes vertical, generating the magnetic field, and are shown by colored arrows in the region on the place of the overlapping of the beams. From the plane, the third axis is similar to the horizontal axis in the plane. The physicists' convention is that in the horizontal plane, the vertical beams are right circularly polarized, whereas the laser beams are left-circularly polarized.

(b) The z-axis is with the local magnetic field direction regardless of the displacement. The model atom possesses a ground state for F = 1 along with no Zeeman splitting, whereas the F = 2 excited state possesses a Zeeman splitting which gives rise to displacement from the center. The energy of the photons is shown with a red arrow. Those that have angular momentum with −z drive transitions that are nearer to resonance, thus, are supposed to be absorbed.

5.4.2. Difficulties in Making MOTs for Molecules

The Doppler cooling in the MOT and the Doppler cooling in free space is the same. A closed cycling transition is a chief requirement, which is still challenging and can be designed for some molecules, as explained in section 3. However, there are more challenges in manufacturing an MOT. They are illustrated by specifying two cases, regarding molecules, where magneto-optical trapping is failed (Tarbutt, 2015).

Refers to Figure 5.8, where it is considered that how an MOT operates for an atom, and the excited state has a Zeeman splitting. Then a minor change is applied, which is shown in Figure 5.9(a) that is a Zeeman splitting applies to the ground state, and not the excited state. Similar to previously, one laser beam causes upward transitions that decrease *M*, whereas the opposing beam causes transitions that grow *M*. However, unlike before, there is an identical detuning of the light for these both transitions. Any preference is not made regarding the absorption of the photons from one beam or the other by the Zeeman splitting of the ground state, therefore, there cannot be a time-averaged confining force. The general convention is that a Zeeman splitting is required by the magneto-optical trapping in the excited state. Extending this concept, it is found that when the Zeeman splitting of the excited state is small, then the confining force is fragile in contrast to that of the ground state.

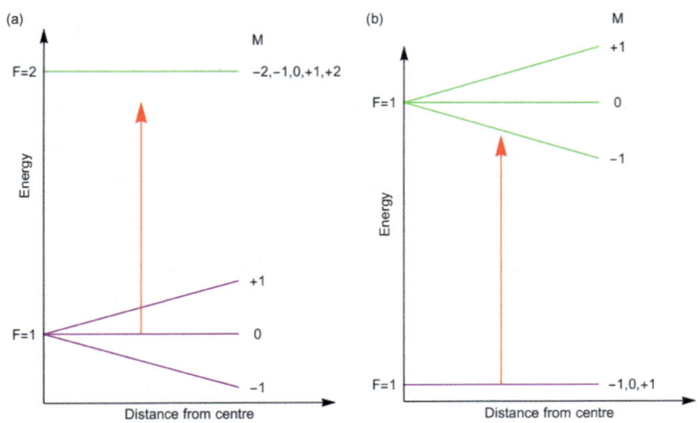

Figure 5.9. Configurations that do not cause MOTs. (a) No Zeeman splitting within the excited state. (b) F = 1 to F = 1 in single dimension.

Source: https://www.sciencedirect.com/topics/earth-and-planetary-sciences/bohr-magneton.

This is exactly the case for all the molecules, where MOTs have been manufactured so far. For example, in the excited state of the CaF, there are two leading contributions to the magnetic moment, one because of the electron spin, and the other because of the orbital angular momentum of the electrons. These revoke nearly exactly, giving up only a little residual magnetic moment. whereas the ground state possesses no orbital angular momentum, hence it has an enormous magnetic moment by the unpaired electron spin. Consequently, there are 40 times smaller Zeeman splitting of the excited state as compared to that in the ground state, in accordance with calculations, it is extremely small to construct a valuable MOT (Tarbutt, 2015).

Consider Figure 5.9(b), where an atom with an $F = 1$ excited state and with an $F = 1$ ground state are shown. There is a Zeeman splitting for the excited state, hence, it might be expected that MOT will work. Assume the happening in one dimension, where there is just a pair of counter-propagating beams. A photon can only be scattered by one of both beams, the one that minimizes M for a ground-state atom in $M = +1$, as there are no levels of the excited state with $M > 1$. Conversely, a photon can only be scattered by the other beam for a ground state atom in $M = -1$, the one that raises M. In both situations, the upper state takes $M = 0$, which decays instinctively to the lower $M = \pm 1$ states having equal probability. Again see Figure 5.8(a), it is seen that an $M = +1$ atom is luminous to the beam that drives towards the center, and overcast to the opposing beam; on the other hand, for an $M = -1$ atom, the contrary is true. The atom is forced to disperse due to the dark state, with just as many photons from one beam as from the other. In addition to that, the net force is zero. In a single dimension, this outcome is particularly for the $F = 1$ to $F = 1$ case. An important role is played by the other beams in three dimensions, and a trapping force is restored. Furthermore, even in a single dimension, there might be a trapping force for alternate angular momentum cases having dark Zeeman levels. However, as compared to cases where there are no dark states, the cases having dark states has a substantially weaker magneto-optical force. This is a challenge for molecules because dark states are always involved by the laser cooling transition as discussed before.

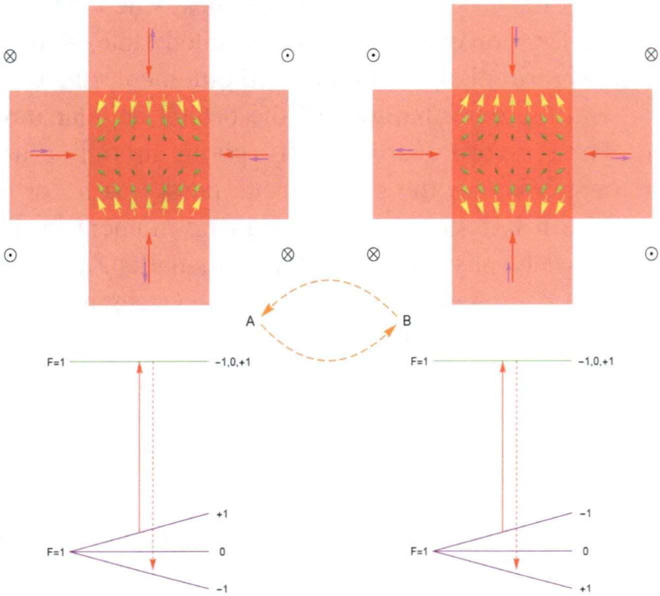

Figure 5.10. Principle of the radio-frequency MOT. The laser polarization and magnetic field are switched back and forth within both configurations demonstrated on the right and left, labeled B and A. In B, molecules are optically pumped back again by the beam, to M = +1, which forces towards the center. In A, molecules are optically pumped by the beam, into M = −1, which drives them towards the center.

Source: https://ui.adsabs.harvard.edu/abs/2019arXiv190205628T.

5.4.3. Radio-Frequency MOT

Assume a molecule with an $F = 1$ excited state and an $F = 1$ ground state, having no Zeeman splitting within the excited state. This is a marvelous model for several molecules that are responsive to laser cooling. It is an example of when a standard MOT will not operate. The principle of a radio-frequency (rf) MOT is shown in Figure 5.10 (Hummon et al., 2013; Norrgard et al., 2016). In this system, the laser polarization handedness and the magnetic field direction both supply backup back and forth within both configurations demonstrated as (labeled A and B). For example, start with configuration A where a molecule moved to the right, and assume the z-axis directed to the right. Due to the polarization of the beam, the photons can be scattered from the beam originating from the right, but not from the beam originating from the left, by a molecule in $M = +1$. Subsequent to the scattering of a few photons, it is presumed to be in $M = -1$, where the circumstances are opposite. But in contrast to $M = +1$, the scattering rate is tedious from $M = -1$ as the transition is farther from resonance having the red-detuned light, hence, the molecule contributes a lot of its time in $M = -1$. After some time, the configuration is transferred all of a sudden from A to B. The molecule lies in the same state, most probably $M = -1$. As the magnetic field has turned around, and also having Zeeman splitting, thus, now this state is the one nearest to resonance. Because the polarizations are turned around as well, once again, the molecule can disperse photons from the beam originating from the right, and not from the beam originating from the left. Shortly, it discovers itself in $M = +1$, and at this time the MOT is diverted back to configuration A, and the cycle iterates. Facilitating the fast enough switching, there is powerful preferential scattering by the beam that drives the molecule backwards from the center. The switching rate must be analogous to the optical pumping rate, generally a few MHz.

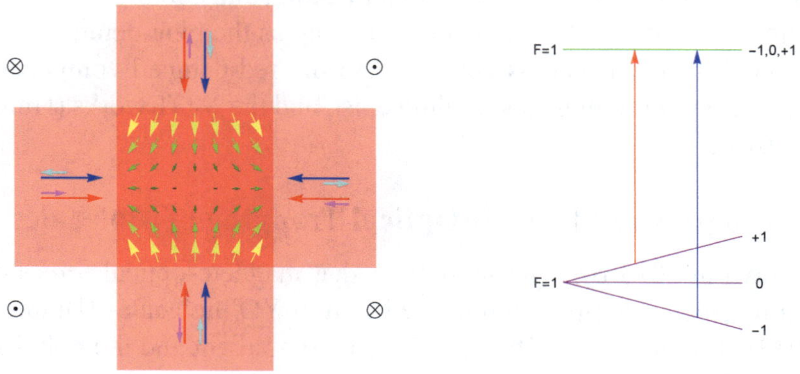

Figure 5.11. Principle of the dual-frequency MOT. Two frequency components involved by each beam, of opposite handedness, one blue-detuned, and the other red-detuned. A molecule in M = −1, from a blue-detuned beam, molecules preferentially scatters photons, again the one that forces it towards the center. When in M = +1, from the red-detuned beam, preferentially scatters photons that drive it towards the center.

Source: https://physics.paperswithcode.com/paper/laser-cooled-molecules.

5.4.4. Dual-Frequency MOT

Another approach for recovering a trapping force is illustrated in Figure 5.11, which is called a dual-frequency MOT (Tarbutt & Steimle, 2015). Hither, two frequency components are involved in each of the MOT beams with adverse handedness, one blue detuned and the other red detuned. Consider a molecule once again that displaced to the right, and consider the *z*-axis parallel to the magnetic field, directed to the right. The photons are scattered preferentially by a molecule in $M = +1$, from a red-detuned element, since this is moved nearer to resonance. Within the two beams, from the right and from the left, the accurate handedness belongs to the one from the right to initiate the transition. And if the molecule is in $M = -1$, it scatters preferentially from a blue-detuned beam, and also it is the beam from the right that possesses the accurate handedness to make this transition happens. Hence, a restoring force is produced by this arrangement towards the hub of the MOT. However, the availability of both blue- and red-detuned light establishes a competition within Doppler heating and cooling. It emerges to the domination of the Doppler cooling as long as the blue-detuned element is farther from resonance as compared to the red-detuned component. In such a case, both trapping and cooling exist, and the MOT works (Hummon et al., 2013).

5.4.5. Progress in Magneto-Optical Trapping of Molecules

It is shown in 2013 by a group at JILA that magneto-optical forces could drive a transverse compression of the beam of YO molecules (Hummon et al., 2013). The direction of the magnetic field gradient and the polarization of the cooling laser are modulated, enforcing the radio-frequency MOT method. Inclusion of the Doppler cooling of the beam to around 2 mK, the compression of the transverse distribution of molecules is also shown when the field direction and polarization are switched in phase, although not when turned out of phase. It has been considered an essential initial first step towards a three-dimensional MOT.

The initial 3D MOT of molecules is demonstrated by the Yale group in 2014 (Barry et al., 2014). where SrF has been used, pursuing on from their previous work on laser cooling and delaying of the alike molecules. A slow beam is produced through optimizing their frequency-broadened slowing method, having a small fraction of the molecules that are inserted into the MOT zone at just a few m/s, which is slow enough to be trapped in the MOT. The rotationally-closed element of the $X(v = 0) \leftrightarrow A(v^0 = 0)$ transition are addressed by the cooling laser. Four hyperfine components of this transition have been used, and the frequency spectrum of the laser are crafted carefully to map each one. Additionally, three more lasers has been used to hinder the leaks to the $v = 3$, $v = 2$ and $v = 1$ vibrational states. From molecules, the laser-induced fluorescence (LIF) trapped in the MOT has been mapped onto a CCD camera. For four cases, images captured from the camera are shown in Figure 5.12.

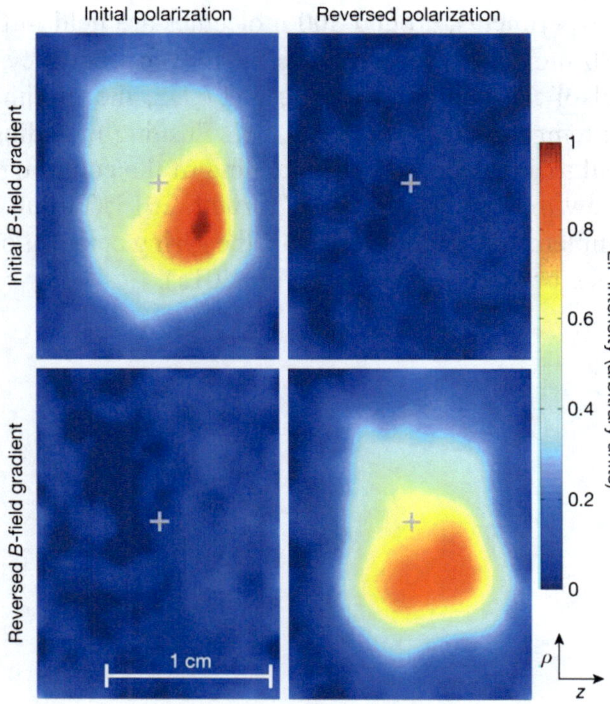

Figure 5.12. Magneto-optical trapping of SrF. The trapped molecules are noticed by capturing their fluorescence with a camera. On the correct choice of the handedness of the laser polarization, the molecules are only trapped, corresponding to the magnetic field direction.

Source: https://www.nature.com/articles/nature13634?page=4.

In the top-left picture, the magnetic field direction and polarizations are arranged to serve a magneto-optical confining force, and a bright area of fluorescence comes out to be close to the region where the magnetic field becomes zero, according to molecules captured around this point. The molecules disappear on the reversal of the magnetic field direction (bottom left), and on the reversal of the polarizations (top right) because of the reversal of the magneto-optical forces which is the reason for the evacuation of the molecules. The molecules show up again on the reversal of both the direction of the magnetic field and the polarization of the light. This order is known to be the smoking gun of an MOT. Moreover, it is also shown that if the molecules are given a short push, they are subjected to a damped oscillation over the trap center, which shows clearly that they are captured. In such initial experiments, around 300 molecules are held in the MOT at a density of 600 molecules per cm^3. The radial trap frequency, a measure of the potential of the confining forces, is 17 Hz, the lifetime has been 56 ms, and the temperature is set to 2.3 mK. Ensuing modeling work has been contributed to an advanced understanding of the confining forces and demonstrates how they could be raised (Tarbutt, 2015; Tarbutt & Steimle, 2015). When implementing these concepts, the spring constant of the trap is increased by a factor of 20 (McCarron et al., 2015).

REFERENCES

1. Abe, M., Prasannaa, V. S., & Das, B. P., (2018). Application of the finite-field coupled-cluster method to calculate molecular properties relevant to electron electric-dipole-moment searches. Physical Review A, 97(3), 032515.
2. Aggarwal, P., Bethlem, H. L., Borschevsky, A., Denis, M., Esajas, K., Haase, P. A., & Zapara, A., (2018). Measuring the electric dipole moment of the electron in BaF. The European Physical Journal D, 72(11), 1–10.
3. Akerman, N., Karpov, M., Segev, Y., Bibelnik, N., Narevicius, J., & Narevicius, E., (2017). Trapping of molecular oxygen together with lithium atoms. Physical Review Letters, 119(7), 073204.
4. Altuntaş, E., Ammon, J., Cahn, S. B., & DeMille, D., (2018). Demonstration of a sensitive method to measure nuclear-spin-dependent parity violation. Physical Review Letters, 120(14), 142501.
5. Anderegg, L., Augenbraun, B. L., Bao, Y., Burchesky, S., Cheuk, L. W., Ketterle, W., & Doyle, J. M., (2018). Laser cooling of optically trapped molecules. Nature Physics, 14(9), 890–893.
6. Anderegg, L., Augenbraun, B. L., Chae, E., Hemmerling, B., Hutzler, N. R., Ravi, A., & Doyle, J. M., (2017). Radiofrequency magneto-optical trapping of CaF with high density. Physical Review Letters, 119(10), 103201.
7. Anderegg, L., Cheuk, L. W., Bao, Y., Burchesky, S., Ketterle, W., Ni, K. K., & Doyle, J. M., (2019). An optical tweezer array of ultracold molecules. Science, 365(6458), 1156–1158.
8. André, A., DeMille, D., Doyle, J. M., Lukin, M. D., Maxwell, S. E., Rabl, P., & Zoller, P., (2006). A coherent all-electrical interface between polar molecules and mesoscopic superconducting resonators. Nature Physics, 2(9), 636–642.
9. Andreev, V., & Hutzler, N. R., (2018). Improved limit on the electric dipole moment of the electron. Nature, 562(7727), 355–360.
10. Augenbraun, B. L., Lasner, Z. D., Frenett, A., Sawaoka, H., Miller, C., Steimle, T. C., & Doyle, J. M., (2020). Laser-cooled polyatomic molecules for improved electron electric dipole moment searches. New Journal of Physics, 22(2), 022003.
11. Barnett, R., Petrov, D., Lukin, M., & Demler, E., (2006). Quantum magnetism with multicomponent dipolar molecules in an optical lattice. Physical Review Letters, 96(19), 190401.

12. Baron, J., Campbell, W. C., DeMille, D., Doyle, J. M., Gabrielse, G., Gurevich, Y. V., & ACME Collaboration, (2014). Order of magnitude smaller limit on the electric dipole moment of the electron. Science, 343(6168), 269–272.
13. Barredo, D., De Léséleuc, S., Lienhard, V., Lahaye, T., & Browaeys, A., (2016). An atom-by-atom assembler of defect-free arbitrary two-dimensional atomic arrays. Science, 354(6315), 1021–1023.
14. Barry, J. F., McCarron, D. J., Norrgard, E. B., Steinecker, M. H., & DeMille, D., (2014). Magneto-optical trapping of a diatomic molecule. Nature, 512(7514), 286–289.
15. Barry, J. F., Shuman, E. S., Norrgard, E. B., & DeMille, D., (2012). Laser radiation pressure slowing of a molecular beam. Physical Review Letters, 108(10), 103002.
16. Bernien, H., Schwartz, S., Keesling, A., Levine, H., Omran, A., Pichler, H., & Lukin, M. D., (2017). Probing many-body dynamics on a 51-atom quantum simulator. Nature, 551(7682), 579–584.
17. Blackmore, J. A., Caldwell, L., Gregory, P. D., Bridge, E. M., Sawant, R., Aldegunde, J., & Cornish, S. L., (2018). Ultracold molecules for quantum simulation: Rotational coherences in CaF and RbCs. Quantum Science and Technology, 4(1), 014010.
18. Cairncross, W. B., & Ye, J., (2019). Atoms and molecules in the search for time-reversal symmetry violation. Nature Reviews Physics, 1(8), 510–521.
19. Cairncross, W. B., Gresh, D. N., Grau, M., Cossel, K. C., Roussy, T. S., Ni, Y., & Cornell, E. A., (2017). Precision measurement of the electron's electric dipole moment using trapped molecular ions. Physical Review Letters, 119(15), 153001.
20. Caldwell, L., Devlin, J. A., Williams, H. J., Fitch, N. J., Hinds, E. A., Sauer, B. E., & Tarbutt, M. R., (2019). Deep laser cooling and efficient magnetic compression of molecules. Physical Review Letters, 123(3), 033202.
21. Cheng, C., Van, D. P. A. P., Jansen, P., Quintero-Pérez, M., Wall, T. E., Ubachs, W., & Bethlem, H. L., (2016). Molecular fountain. Physical Review Letters, 117(25), 253201.
22. Cheuk, L. W., Anderegg, L., Augenbraun, B. L., Bao, Y., Burchesky, S., Ketterle, W., & Doyle, J. M., (2018). Λ-enhanced imaging of molecules in an optical trap. Physical Review Letters, 121(8), 083201.

23. Cocks, D. G., & Whittingham, I. B., (2009). Laser-intensity dependence of photoassociation in ultracold metastable helium. Physical Review A, 80(2), 023417.

24. Cocks, D. G., & Whittingham, I. B., (2010). Photoassociation spectra and the validity of the dipole approximation for weakly bound dimers. Physical Review A, 81(3), 033406.

25. Cocks, D. G., Whittingham, I. B., & Peach, G., (2010). Effects of non-adiabatic and Coriolis couplings on the bound states of He (2 ^{3}S)+ He (2 ^{3}P). Journal of Physics B: Atomic, Molecular and Optical Physics, 43(13), 135102.

26. Collopy, A. L., Ding, S., Wu, Y., Finneran, I. A., Anderegg, L., Augenbraun, B. L., & Ye, J., (2018). 3D magneto-optical trap of yttrium monoxide. Physical Review Letters, 121(21), 213201.

27. Collopy, A. L., Hummon, M. T., Yeo, M., Yan, B., & Ye, J., (2015). Prospects for a narrow line MOT in YO. New Journal of Physics, 17(5), 055008.

28. Dalibard, J., & Cohen-Tannoudji, C., (1989). Laser cooling below the Doppler limit by polarization gradients: simple theoretical models. JOSA B, 6(11), 2023–2045.

29. Dall, R. G., Dedman, C. J., & Truscott, A. G., (2008). Feedback control of an atom laser. Optics Express, 16(19), 14716–14722.

30. De Marco, L., Valtolina, G., Matsuda, K., Tobias, W. G., Covey, J. P., & Ye, J., (2019). A degenerate fermi gas of polar molecules. Science, 363(6429), 853–856.

31. Deguilhem, B., Leininger, T., Gadéa, F. X., & Dickinson, A. S., (2008). Photo association of He2 revisited. Journal of Physics B: Atomic, Molecular and Optical Physics, 42(1), 015102.

32. DeMille, D., (2002). Quantum computation with trapped polar molecules. Physical Review Letters, 88(6), 067901.

33. DeMille, D., Doyle, J. M., & Sushkov, A. O., (2017). Probing the frontiers of particle physics with tabletop-scale experiments. Science, 357(6355), 990–994.

34. Devlin, J. A., & Tarbutt, M. R., (2016). Three-dimensional Doppler, polarization-gradient, and magneto-optical forces for atoms and molecules with dark states. New Journal of Physics, 18(12), 123017.

35. Di Rosa, M. D., (2004). Laser-cooling molecules. The European Physical Journal D-Atomic, Molecular, Optical and Plasma Physics, 31(2), 395–402.

36. Ding, S., Wu, Y., Finneran, I. A., Burau, J. J., & Ye, J., (2020). Sub-doppler cooling and compressed trapping of YO molecules at μ K temperatures. Physical Review X, 10(2), 021049.
37. Fitch, N. J., & Tarbutt, M. R., (2016). Principles and design of a Zeeman-Sisyphus decelerator for molecular beams. ChemPhysChem, 17(22), 3609.
38. Galica, S. E., Aldridge, L., McCarron, D. J., Eyler, E. E., & Gould, P. L., (2018). Deflection of a molecular beam using the bichromatic stimulated force. Physical Review A, 98(2), 023408.
39. Gorshkov, A. V., Manmana, S. R., Chen, G., Ye, J., Demler, E., Lukin, M. D., & Rey, A. M., (2011). Tunable superfluidity and quantum magnetism with ultracold polar molecules. Physical Review Letters, 107(11), 115301.
40. Hao, Y., Pašteka, L. F., Visscher, L., Aggarwal, P., Bethlem, H. L., Boeschoten, A., & (NL-e EDM Collaboration), (2019). High accuracy theoretical investigations of CaF, SrF, and BaF and implications for laser-cooling. The Journal of Chemical Physics, 151(3), 034302.
41. Hemmerling, B., Chae, E., Ravi, A., Anderegg, L., Drayna, G. K., Hutzler, N. R., & Doyle, J. M., (2016). Laser slowing of CaF molecules to near the capture velocity of a molecular MOT. Journal of Physics B: Atomic, Molecular and Optical Physics, 49(17), 174001.
42. Ho, C. J., Devlin, J. A., Rabey, I. M., Yzombard, P., Lim, J., Wright, S. C., & Sauer, B. E., (2020). New techniques for a measurement of the electron's electric dipole moment. New Journal of Physics, 22(5), 053031.
43. Hudson, E. R., Lewandowski, H. J., Sawyer, B. C., & Ye, J., (2006). Cold molecule spectroscopy for constraining the evolution of the fine structure constant. Physical Review Letters, 96(14), 143004.
44. Hudson, J. J., Kara, D. M., Smallman, I. J., Sauer, B. E., Tarbutt, M. R., & Hinds, E. A., (2011). Improved measurement of the shape of the electron. Nature, 473(7348), 493–496.
45. Hummon, M. T., Yeo, M., Stuhl, B. K., Collopy, A. L., Xia, Y., & Ye, J., (2013). 2D magneto-optical trapping of diatomic molecules. Physical Review Letters, 110(14), 143001.
46. Isaev, T. A., & Berger, R., (2016). Polyatomic candidates for cooling of molecules with lasers from simple theoretical concepts. Physical Review Letters, 116(6), 063006.

47. Jung, R., Gerlach, S., Schumann, R., Von, O. G., & Eichmann, U., (2003). Magneto-optical trapping of stark-slowed metastable He atoms. The European Physical Journal D-Atomic, Molecular, Optical and Plasma Physics, 23(3), 415–419.
48. Kaufman, A. M., Lester, B. J., & Regal, C. A., (2012). Cooling a single atom in an optical tweezer to its quantum ground state. Physical Review X, 2(4), 041014.
49. Kozyryev, I., & Hutzler, N. R., (2017). Precision measurement of time-reversal symmetry violation with laser-cooled polyatomic molecules. Physical Review Letters, 119(13), 133002.
50. Kozyryev, I., Baum, L., Aldridge, L., Yu, P., Eyler, E. E., & Doyle, J. M., (2018). Coherent bichromatic force deflection of molecules. Physical Review Letters, 120(6), 063205.
51. Kozyryev, I., Baum, L., Matsuda, K., Augenbraun, B. L., Anderegg, L., Sedlack, A. P., & Doyle, J. M., (2017). Sisyphus laser cooling of a polyatomic molecule. Physical Review Letters, 118(17), 173201.
52. Krems, R. V., (2008). Cold controlled chemistry. Physical Chemistry Chemical Physics, 10(28), 4079–4092.
53. Lim, J., Almond, J. R., Trigatzis, M. A., Devlin, J. A., Fitch, N. J., Sauer, B. E., & Hinds, E. A., (2018). Laser cooled YbF molecules for measuring the electron's electric dipole moment. Physical Review Letters, 120(12), 123201.
54. Lim, J., Frye, M. D., Hutson, J. M., & Tarbutt, M. R., (2015). Modeling sympathetic cooling of molecules by ultracold atoms. Physical Review A, 92(5), 053419.
55. Liu, L. R., Hood, J. D., Yu, Y., Zhang, J. T., Hutzler, N. R., Rosenband, T., & Ni, K. K., (2018). Building one molecule from a reservoir of two atoms. Science, 360(6391), 900–903.
56. McCarron, D. J., Norrgard, E. B., Steinecker, M. H., & DeMille, D., (2015). Improved magneto-optical trapping of a diatomic molecule. New Journal of Physics, 17(3), 035014.
57. McCarron, D. J., Steinecker, M. H., Zhu, Y., & DeMille, D., (2018). Magnetic trapping of an ultracold gas of polar molecules. Physical Review Letters, 121(1), 013202.
58. Micheli, A., Brennen, G. K., & Zoller, P., (2006). A toolbox for lattice-spin models with polar molecules. Nature Physics, 2(5), 341–347.

59. Morita, M., Kosicki, M. B., Żuchowski, P. S., & Tscherbul, T. V., (2018). Atom-molecule collisions, spin relaxation, and sympathetic cooling in an ultracold spin-polarized Rb (S 2)− SrF (Σ+ 2) mixture. Physical Review A, 98(4), 042702.

60. Moses, S. A., Covey, J. P., Miecnikowski, M. T., Jin, D. S., & Ye, J., (2017). New frontiers for quantum gases of polar molecules. Nature Physics, 13(1), 13–20.

61. Nayak, M. K., & Chaudhuri, R. K., (2006). Ab initio calculation of P, T-odd interaction constant in BaF: A restricted active space configuration interaction approach. Journal of Physics B: Atomic, Molecular and Optical Physics, 39(5), 1231.

62. Norrgard, E. B., Edwards, E. R., McCarron, D. J., Steinecker, M. H., DeMille, D., Alam, S. S., & Hunter, L. R., (2017). Hyperfine structure of the B 3 Π 1 state and predictions of optical cycling behavior in the X→ B transition of TlF. Physical Review A, 95(6), 062506.

63. Norrgard, E. B., Edwards, E. R., McCarron, D. J., Steinecker, M. H., DeMille, D., Alam, S. S., & Hunter, L. R., (2017). Hyperfine Structure of the $B^3\pi_1$ State and Predictions of Optical Cycling Behavior in the X->B Transition of TlF. arXiv preprint arXiv:1702.02548.

64. Norrgard, E. B., McCarron, D. J., Steinecker, M. H., Tarbutt, M. R., & DeMille, D., (2016). Sub-milli-Kelvin dipolar molecules in a radio-frequency magneto-optical trap. Physical Review Letters, 116(6), 063004.

65. Parazzoli, L. P., Fitch, N. J., Żuchowski, P. S., Hutson, J. M., & Lewandowski, H. J., (2011). Large effects of electric fields on atom-molecule collisions at milli-Kelvin temperatures. Physical Review Letters, 106(19), 193201.

66. Park, J. W., Yan, Z. Z., Loh, H., Will, S. A., & Zwierlein, M. W., (2017). Second-scale nuclear spin coherence time of ultracold 23Na40K molecules. Science, 357(6349), 372–375.

67. Phillips, W. D., (1998). Nobel Lecture: Laser cooling and trapping of neutral atoms. Reviews of Modern Physics, 70(3), 721.

68. Prehn, A., Ibrügger, M., Glöckner, R., Rempe, G., & Zeppenfeld, M., (2016). Optoelectrical cooling of polar molecules to submilli-Kelvin temperatures. Physical Review Letters, 116(6), 063005.

69. Reens, D., Wu, H., Langen, T., & Ye, J., (2017). Controlling spin flips of molecules in an electromagnetic trap. Physical Review A, 96(6), 063420.

70. Richter, F., Becker, D., Bény, C., Schulze, T. A., Ospelkaus, S., & Osborne, T. J., (2015). Ultracold chemistry and its reaction kinetics. New Journal of Physics, 17(5), 055005.
71. Safronova, M. S., Budker, D., DeMille, D., Kimball, D. F. J., Derevianko, A., & Clark, C. W., (2018). Search for new physics with atoms and molecules. Reviews of Modern Physics, 90(2), 025008.
72. Schlosser, N., Reymond, G., Protsenko, I., & Grangier, P., (2001). Sub-Poissonian loading of single atoms in a microscopic dipole trap. Nature, 411(6841), 1024–1027.
73. Shelkovnikov, A., Butcher, R. J., Chardonnet, C., & Amy-Klein, A., (2008). Stability of the proton-to-electron mass ratio. Physical Review Letters, 100(15), 150801.
74. Shuman, E. S., Barry, J. F., & DeMille, D., (2010). Laser cooling of a diatomic molecule. Nature, 467(7317), 820–823.
75. Shuman, E. S., Barry, J. F., Glenn, D. R., & DeMille, D., (2009). Radiative force from optical cycling on a diatomic molecule. Physical Review Letters, 103(22), 223001.
76. Steinecker, M. H., McCarron, D. J., Zhu, Y., & DeMille, D., (2016). Improved radio-frequency magneto-optical trap of SrF molecules. ChemPhysChem, 17(22), 3664–3669.
77. Stuhl, B. K., Hummon, M. T., Yeo, M., Quéméner, G., Bohn, J. L., & Ye, J., (2012). Evaporative cooling of the dipolar hydroxyl radical. Nature, 492(7429), 396–400.
78. Stuhl, B. K., Sawyer, B. C., Wang, D., & Ye, J., (2008). Magneto-optical trap for polar molecules. Physical Review Letters, 101(24), 243002.
79. Tarbutt, M. R., & Steimle, T. C., (2015). Modeling magneto-optical trapping of CaF molecules. Physical Review A, 92(5), 053401.
80. Tarbutt, M. R., (2015). Magneto-optical trapping forces for atoms and molecules with complex level structures. New Journal of Physics, 17(1), 015007.
81. Tarbutt, M. R., Hudson, J. J., Sauer, B. E., & Hinds, E. A., (2009). Prospects for measuring the electric dipole moment of the electron using electrically trapped polar molecules. Faraday Discussions, 142, 37–56.
82. Tarbutt, M. R., Sauer, B. E., Hudson, J. J., & Hinds, E. A., (2013). Design for a fountain of YbF molecules to measure the electron's electric dipole moment. New Journal of Physics, 15(5), 053034.

83. Tokunaga, S. K., Skomorowski, W., Żuchowski, P. S., Moszynski, R., Hutson, J. M., Hinds, E. A., & Tarbutt, M. R., (2011). Prospects for sympathetic cooling of molecules in electrostatic, ac and microwave traps. The European Physical Journal D, 65(1), 141–149.

84. Tokunaga, S. K., Stoeffler, C., Auguste, F., Shelkovnikov, A., Daussy, C., Amy-Klein, A., & Darquié, B., (2013). Probing weak force-induced parity violation by high-resolution mid-infrared molecular spectroscopy. Molecular Physics, 111(14, 15), 2363–2373.

85. Truppe, S., Hendricks, R. J., Tokunaga, S. K., Lewandowski, H. J., Kozlov, M. G., Henkel, C., & Tarbutt, M. R., (2013). A search for varying fundamental constants using hertz-level frequency measurements of cold CH molecules. Nature Communications, 4(1), 1–7.

86. Truppe, S., Marx, S., Kray, S., Doppelbauer, M., Hofsäss, S., Schewe, H. C., & Meijer, G., (2019). Spectroscopic characterization of aluminum monofluoride with relevance to laser cooling and trapping. Physical Review A, 100(5), 052513.

87. Truppe, S., Williams, H. J., Fitch, N. J., Hambach, M., Wall, T. E., Hinds, E. A., & Tarbutt, M. R., (2017). An intense, cold, velocity-controlled molecular beam by frequency-chirped laser slowing. New Journal of Physics, 19(2), 022001.

88. Truppe, S., Williams, H. J., Hambach, M., Caldwell, L., Fitch, N. J., Hinds, E. A., & Tarbutt, M. R., (2017). Molecules cooled below the Doppler limit. Nature Physics, 13(12), 1173–1176.

89. Tscherbul, T. V., Kłos, J., & Buchachenko, A. A., (2011). Ultracold spin-polarized mixtures of $^2\Sigma$ molecules with S-state atoms: Collisional stability and implications for sympathetic cooling. Physical Review A, 84(4), 040701.

90. Ungar, P. J., Weiss, D. S., Riis, E., & Chu, S., (1989). Optical molasses and multilevel atoms: Theory. JOSA B, 6(11), 2058–2071.

91. Van De, M. S. Y., Bethlem, H. L., Vanhaecke, N., & Meijer, G., (2012). Manipulation and control of molecular beams. Chemical Reviews, 112(9), 4828–4878.

92. Weidemüller, M., Esslinger, T., Ol'shanii, M. A., Hemmerich, A., & Hänsch, T. W., (1994). A novel scheme for efficient cooling below the photon recoil limit. EPL (Europhysics Letters), 27(2), 109.

93. Williams, H. J., Caldwell, L., Fitch, N. J., Truppe, S., Rodewald, J., Hinds, E. A., & Tarbutt, M. R., (2018). Magnetic trapping and coherent control of laser-cooled molecules. Physical Review Letters, 120(16), 163201.

94. Williams, H. J., Truppe, S., Hambach, M., Caldwell, L., Fitch, N. J., Hinds, E. A., & Tarbutt, M. R., (2017). Characteristics of a magneto-optical trap of molecules. New Journal of Physics, 19(11), 113035.
95. Wu, X., Gantner, T., Koller, M., Zeppenfeld, M., Chervenkov, S., & Rempe, G., (2017). A cryofuge for cold-collision experiments with slow polar molecules. Science, 358(6363), 645–648.
96. Yan, B., Moses, S. A., Gadway, B., Covey, J. P., Hazzard, K. R., Rey, A. M., & Ye, J., (2013). Observation of Dipolar Spin-exchange Interactions with Polar Molecules in a Lattice. Colorado Univ at boulder dept of physics.
97. Yelin, S. F., Kirby, K., & Côté, R., (2006). Schemes for robust quantum computation with polar molecules. Physical Review A, 74(5), 050301.
98. Yeo, M., Hummon, M. T., Collopy, A. L., Yan, B., Hemmerling, B., Chae, E., & Ye, J., (2015). Rotational state microwave mixing for laser cooling of complex diatomic molecules. Physical Review Letters, 114(22), 223003.
99. Zhelyazkova, V., Cournol, A., Wall, T. E., Matsushima, A., Hudson, J. J., Hinds, E. A., & Sauer, B. E., (2014). Laser cooling and slowing of CaF molecules. Physical Review A, 89(5), 053416.
100. Zhou, Y., Shagam, Y., Cairncross, W. B., Ng, K. B., Roussy, T. S., Grogan, T., & Cornell, E. A., (2020). Second-scale coherence measured at the quantum projection noise limit with hundreds of molecular ions. Physical Review Letters, 124(5), 053201.

Chapter

6

Interaction between Cold Atoms and Cold Molecules in an Electrostatic-Magnetic Trap

CONTENTS

6.1. INTRODUCTION

Interactions measurement among ultracold atoms and cold molecules can be studied in depth to gain a better understanding of basic collision processes. These measurements can be made in a variety of experimental geometries, such as when a beam has both species, where one species is stuck, or when both species are stuck. Concurrent trapping allows for slightly longer contact times and, as a result, a higher sensitivity to unusual collisions (Collyer & Clegg, 1986). However, integrating molecular and atomic structures, which also have competing experimental criteria, poses major functional challenges. In this study, we explain in detail an experimental method for studying cold collisions among cold molecules and ultracold atoms in a dual trap, in which the molecules and atoms are trapped utilizing static electric and magnetic fields simultaneously. We investigate cold collisions among rubidium (87Rb and 85Rb) atoms and ammonia (14ND3 and 15ND3) molecules as a description of the system's abilities (Malzahn & Schultz, 1986; Larsen, 2000).

The research of interactions concerning ultracold neutral or cold molecules is a hot topic in science right now (Kadolph & Casselman, 2004; Chen et al., 2008). This research aims to understand the basic quantum nature of interactions by studying inelastic and elastic scattering, along with chemical reactions. Recent achievements include the discovery of light-assisted collisions in optical tweezers of laser-cooled CaF molecules and the use of stereodynamic effects to regulate chemical reactions within ultracold KRb molecules reactions among excited metastable noble gas molecules and atoms probing during Penning ionization (PI), of quantum scattering resonances analyzes of reactions between polar molecules and dipolar collisions as well as in crossed molecular beams of NO and H2 the measurements of high-resolution differential cross-sections. Experimentally evaluating collisions among trapped molecules and atoms, such as Li + O2 N + NH and Rb + ND3 are particularly relevant to the work addressed here (Nguyen-Chung & Mennig, 2001; Senthilvelan & Gnanamoorthy, 2006).

Based on the species in the study, temperature regimes, and research objectives, different experimental architectures are being used to investigate interactions (Patel & Padhi, 1991). Experiments with cold molecules usually use fused or crossed molecular beams, a beam of molecules impinging on a stuck sample, or a system in which both species are stuck. The latter experiments are either cotraps, in which the constraining force for both species is produced by the same field (magnetic, electric, or optical) or dual traps, in which the constraining force for each species is generated by a different field. Each platform has its own set of benefits and drawbacks. High-resolution collision energy tuning is possible with fused or crossed beams (Nguyen-Chung et al., 1998; Barbosa & Kenny, 1999). Collisions at lower temperatures can also be studied in the case of fused beams (Belgacem et al., 1994; Yadav et al., 2021). However, beam-based experiments have limited contact periods and are less sensitive to occasional collision events. For nonmerged beam tests, the lower temperatures that can be probed are often limited by the forward speed of the minimum beam.

Trapping all species allows for long-interaction-time and low-energy experiments, enabling them for higher sensitivity even for the rarest collision events. Furthermore, elastic and inelastic collisions can be separated, with the former causing an ant trapped state that can be calculated as trap failure, and the latter causing thermalization among the two species. Sympathetic cooling is possible at the absolute end of elastic collisions controlling interactions (Kaplan & Hansen, 1997). This last probability allows trapped collisions especially appealing as a way to create ultracold molecules utilizing well-established ultracold atom techniques without any necessity for molecular laser cooling or being limited to bialkali species (Park et al., 2002; Yue & Padmanabhan, 1999). Sadly, theoretical calculations utilizing potential energy surfaces (PESs) of ab initio are always insufficient to determine whether a given atom-molecular structure would have a high adequate ratio of elastic to inelastic collision cross sections to enable sympathetic cooling. To constrain theoretical models, direct experimental measurements are required. Experimenting with various isotopologs is also important for limiting ab initio forecasting (Wu et al., 2004).

Despite the benefits, there are a variety of problems with the use of a trap to investigate collisions. First, since the events of inelastic scattering occur over a long timescale and the corresponding products are normally not trapped, measuring the product's outgoing or the states of quantum directly is generally impractical. As a result, inelastic collisions can only be observed implicitly from trap failure calculations. We should point out that the latest research on bialkali reactions an unusual and thrilling are exceptions to this rule (Chung & Kwon, 2002). Secondly, differential cross sections cannot be measured in a trap due to the complex particle dynamics. Third, the distribution of density in the trap is not constant in time or space, causing the collision rates to be temporally and spatially dependent. Finally, if somehow the dynamics of collision are influenced by external fields, the trapping fields themselves, which are spatially inhomogeneous, will create additional spatial dependency. The extraction of articulated cross-sections is difficult due to these complications. Despite these difficulties, the possible advantages of using a trap to research collisions make them worthwhile (Barbosa & Kenny, 1999; Fu et al., 1999).

6.2. EXPERIMENTAL DESCRIPTION FOR INTERACTIONS OF ATOMIC INTERACTIONS AND COLD MOLECULAR

Figure 6.1 shows a schematic diagram of the laboratory apparatus (top). Cooling lasers, cylindrical glass cells, and anti-Helmholtz coils (black discs) make up the ultracold atom system.

The cold molecule system consists of a pulsed valve (also not shown), and an electrostatic trap formed by four electrodes mounted to the end of the decelerator, A pulsed valve, and a stark decelerator (alternating white/black rod pairs) make up the cold molecule device.

An all-valve of metal gate fixed within the molecule and atom systems enables the two components of the experiment to be baked independently or vented (Gupta & Wang, 1993; Patcharaphun & Mennig, 2005). The dual trap is made by a first filling magneto-optical trap (MOT) in Rb magnetic trap and then the coil pair is translated along the y axis so that the atoms trapped are located at the middle of the electrodes of the electrostatic trap, as seen in Figure 6.1 (bottom).

The Rb atoms cloud will reach the dual-trap area through a 5 mm gap between the two electrodes located centrally. If the atoms are in position, the environment of dual-trap is formed by creating a beam of ND3 (pulsed molecular supersonic), decelerating it, and trapping it in the same physical position. Multiphoton ionization (use for molecules) and absorption imaging (use for atoms) are used to characterize all trapped species before and after a regulated contact time (for both species).

Ions are determined by biasing the electrodes of the electrostatic trap so that ions are collected into the mass spectrometer of time-of-flight with a detector (two-stage microchannel plate (MCP)), which can be seen next to the gate valve as two gray discs (Andrews & Peters, 1998).

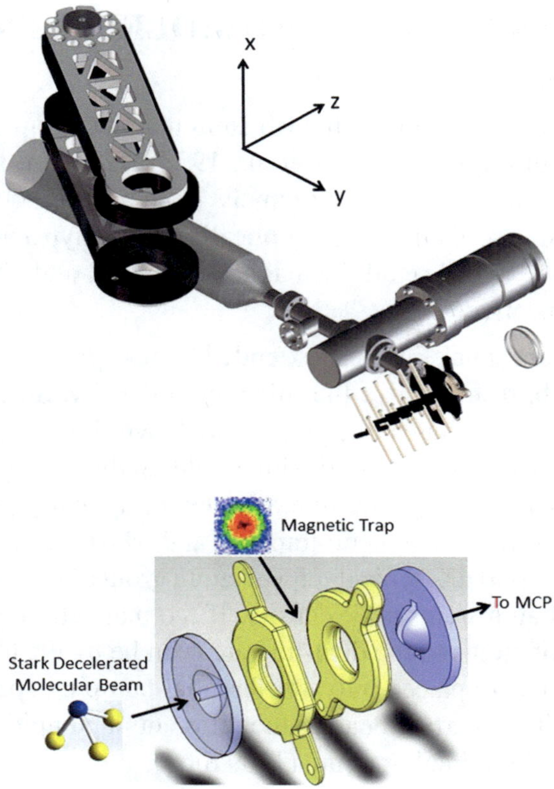

Figure 6.1. Description of the experimental configuration for the dual trap (Not to scale)

Source: https://www.x-mol.com/paper/1368348589426368512?recommendPaper=1280616849208950784.

Note: (Top) The electrostatic trap position, is in the bottom right-hand corner, where the dual trap is built, at the end of a Stark decelerator (alternating rod pairs). External to the Rb vapor cell, a magnetic trap is created by a translatable coil pair (dark discs placed on long arms). (Bottom) Close-up of the dual-trap area, revealing the four electrodes that create the molecules' electrostatic trap. For clarification, the spacing among the electrodes has been enhanced.

6.3. DEVELOPING TRAPPED MOLECULES AND ATOMS

The general form for generating the Rb atom magnetic trap was previously explored in detail (Nguyen-Chung et al., 1998). The first step is to make a MOT out of one of the two extremely popular Rb isotopes, 85Rb or 87Rb. The atoms are then injected optically into a hyperfine state called |F, MF, where MF and F stand for magnetic sublevel and absolute angular momentum, respectively.

The atoms are stuck in their extended states, |*F*, *MF* for 85Rb, and |*F*, *MF* for 87Rb, to maximize the filled atom density. Since this analysis just looks at these two states, the |F, MF labels will be considered for each isotope. The atoms are transported with a field gradient of 375 G/cm in the high dimension into a magnetic trap after being optically pumped. Figure 6.1(top) shows how the magnetic trapping and MOT fields are formed by the same coils. At 600 μK, typical trapped atom groups consist of 109 atoms, leading to 1010 atoms/cm^3 peak density. If the magnetic trap is filled, the transformation of atoms to the dual-trap area can be by the track translation on which the coils are placed into a differential pumping aperture (Akay & Barkley, 1991). The aperture enables the differential pumping and the MOT chamber's needed minimal pressure (1109 torrs).

6.4. CHARACTERIZATION OF TRAPPED POPULATION

To consider the dynamics of all species in the dual trap, a variety of techniques are used, namely ionization (for both molecules and atoms) and for atom absorption detection methods. The approaches in combination permit for exact measurement of the spatial distribution and population time dependence for both species in the dual trap. The two individually trapped populations are both aligned using the same dimensions (Chung & Kwon, 2002).

A 2+1 resonantly improved multiphoton ionization (REMPI) scheme operating close to 317 nm is used to ascertain the trapped molecular sample properties (Margrave & Lagow, 1979). Pumped a pulsed dye laser by 532-nm light from a doubled Nd: YAG serves as the detector laser. A 50-cm lens focuses the 10-mJ output pulse into the trap area with a 20μm beam waist.

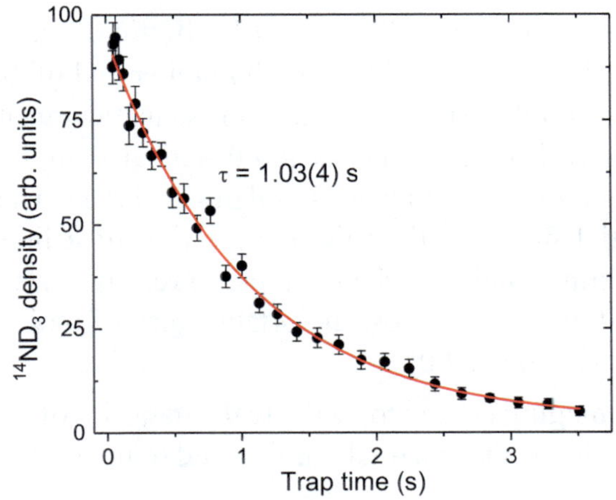

Figure 6.2. $^{14}ND_3$ density at the trap core as a function of time trapping (points). $^{15}ND_3$ caught in traps have equal densities and trap lifetimes. A fit to a single exponential decay with a time constant is represented by the line.

Source: https://jila.colorado.edu/lewandowski/publications/collisions-between-ultracold-atoms-and-cold-molecules-dual-electrostatic-magnetic-trap.

A few microseconds before activating the ionization beam, the electrodes of the electrostatic trap are quickly switched into a 1,1,0,0 kV configuration, extracting the resulting ions. The ions are accelerated onto the MCP by the resulting electric field, which effectively makes the electrostatic trap a time-of-flight mass spectrometer. Because the detection region's REMPI laser beam waist is significantly smaller. So, by scanning the laser position, the typical trapped molecular cloud width (1 mm), the vertical (x) and horizontal (z) profiles of the molecular cloud can be evaluated (Zhu et al., 2017). A slit in the last of the four trapping electrodes, measuring 2mm wide and 1cm tall, allows ions to pass through (this electrode) to the MCP as illustrated in Figure 6.1(bottom). Using the trapping potentials and measurements of the spatial distribution of the molecules in the trap, a rough estimate of the temperature or energy distribution of the sample can be made. Thus, considering the absence of thermalizing elastic collisions in the low-density sample, the cloud is not in thermal equilibrium (Theocaris & Papanicolaou, 1980; Yadav et al., 2021).

The ND3 density at the trap center as a function of trapping time for 14ND3 can be measured as illustrated in Figure 6.2, using the REMPI laser. The experimental measurements are represented as black dots with statistical (1σ) error bars. While a single exponential fit is represented by a solid line. According to the fit, the in-trap $1/e$ lifetime is about one second. The data for 15ND3 is comparable in both number and lifetime. Molecules are lost from traps for three main reasons: nonadiabatic transitions at the trap center collisions with background gas, and optical pumping to untrapped states by blackbody radiation (Jagur-Grodzinski, 1992). The quality of the single exponential fit, as well as the observation of a linear relationship between the lifetime and the chamber pressure, suggests that collisions with background gas are the most important cause of molecule loss in our experiment (Kharitonov, 2000).

Absorption imaging is used to evaluate the majority of characteristics of the stranded Rb cloud, such as total number and temperature.

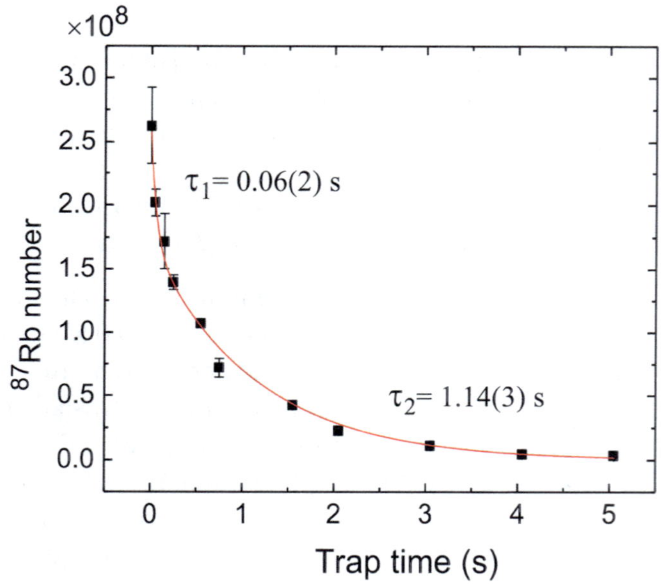

Figure 6.3. Upon being exposed to the fields of electrostatic trapping for various periods of time, the number of 87Rb atoms (black squares) was counted. The decay observed is easily explained by a double exponential form (line). The initial quick decay is caused by the loss of atoms caused by the electric field's reduction of the trapping potential. Collisions with background gas cause the gradual decay.

Source: https://www.researchgate.net/publication/339967170_Collisions_between_ultracold_atoms_and_cold_molecules_in_a_dual_electrostatic-magnetic_trap.

Absorption images are obtained at two separate physical locations, one at the dual trap location and the other 20 cm away in the MOT cell. Magnetic fields created by eddy currents in the steel vacuum chamber generated when the large magnetic trapping field is rapidly switched off, influence absorption images taken in the dual-trap region. The atoms are Zeeman-shifted out of resonance with the laser probe throughout these currents. Which last only a few milliseconds. For short expansion times, quantitative measurements are therefore unreliable in this region. As a result, absorption images obtained in a dual-trap environment are often used for coarse diagnosis. The MOT cell's secondary absorption image position, which is made of glass, is not affected by eddy currents. As a result, the atoms are moved to this location for all number and temperature measurements. Because no heating of the atom cloud was detected throughout transit, such a procedure was possible (Patcharaphun & Mennig, 2005; Rezaei et al., 2009). When the magnetic trap overlaps with the electrostatic trap, Rb's dc polarizability is expected to result in atom loss. The total Rb number is measured as a function of time using absorption images in the presence of electrostatic trapping fields but not in the presence of ND3, as shown in Figure 6.3. After the Rb is returned to the MOT cell, the absorption measurements are completed. We notice two distinct decay timescales, which we approximate with a double exponential decay of the form.

$$NRb(t) = (N_0 - N_1)e^{-t\left(\frac{\tau_1+\tau_2}{\tau_1\tau_2}\right)} + N_1 e^{-} t/\tau 2, \qquad (1)$$

where $N_{Rb}(t)$ denotes the total number of Rb atoms in the trapped cloud at time t, N_0 denotes the initial number of atoms, and N_1 denotes the number of atoms that remain trapped in the presence of an electrostatic field.

Table 6.1. Rb Number Parameters of decay in the Dual-Trap Field with Trap Electrodes at 8kV

	$N_0\,(10^8)$	$N_1\,(10^8)$	τ_1 (s)	τ_2 (s)
^{87}Rb	2.6(5)	1.7(1)	0.06(2)	1.14(3)
^{85}Rb	1.5(3)	0.8(2)	0.05(2)	0.90(1)

Once the large electrostatic trapping fields are applied, the initial fast decay (τ1) is due to a drop in the trap depth, and due to the dc polarizability of Rb. Once the high-energy atoms have left the trap, this decay mechanism is turned off. Collisions with background gas cause slower decay. The population loss follows a single exponential decay with the length of the two-time constants (τ2) if the electric fields are turned off completely. The behavior of both Rb isotopes in terms of population is summarized in Table

6.1. The experiment is optimized for 87Rb, and the number of 85Rb atoms we could reliably load into the magnetic trap was half that of 87Rb. This explains the two-fold discrepancy in the initial number loaded (Scola & Brooks, 1970; Leon et al., 2005).

6.5. DUAL-TRAP ALIGNMENT

The measurement techniques addressed so far facilitate the geographical alignment between the two trapped populations, in addition to characterising the in-trap dynamics and conduct of both species. This is accomplished through a combination of techniques. Adjustments to the dual-trap overlap are formed by adjusting the position of the magnetic trapping coils because the electrostatic trap placement is fixed, which are situated on the outside of the vacuum chamber. As, the alignment in one aspect is not always orthogonal to the other aspects, so the process of aligning the traps requires adjustments in all three aspects. It is an iterative process. By scanning the ionization laser and overlapping the profiles of the molecule and atom samples, the vertical (x) dimension is synchronized, as shown in Figure 6.4(a). Vertical translation of the coil pair or a slight imbalance in the two independently regulated coil currents could be used to adjust the height of the magnetic trap.

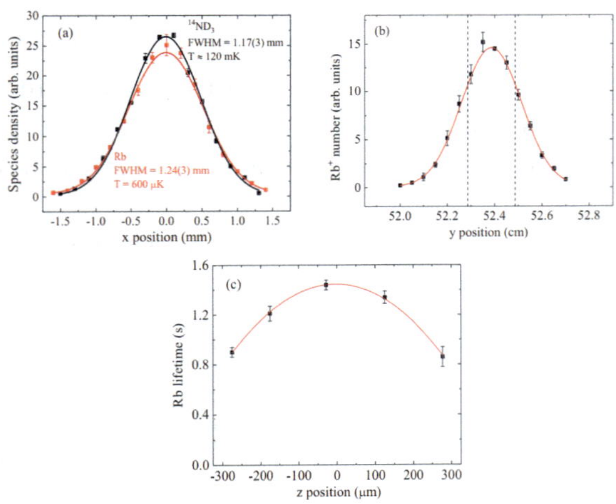

Figure 6.4. Gaussian fits are used to coordinate the two isolated traps in all three spatial dimensions (solid lines).

Source: https://www.researchgate.net/publication/339967170_Collisions_between_ultracold_atoms_and_cold_molecules_in_a_dual_electrostatic-magnetic_trap.

Note: (a) The profile of both ND3 and Rb can be calculated in the x dimension by scanning the ionization laser. The widths of the two data sets have been scaled to make comparisons easier. (b) The ionized Rb atoms number is seen in the y dimension for different final track positions. The vertically dashed lines represent the slit extents in the last electrostatic-trap electrode as viewed. (c) When the trapped atom cloud is oriented to the core of the fields of electrostatic trapping in the z-axis, the Rb lifespan is maximized.

Unexpectedly, the widths of the atom and molecule traps in the vertical axis, are almost identical. This dimension's near-perfect spatial overlap is unique. Thus, the magnetic trap has a strong axis, while the electrostatic trap has a weak axis. The Gaussian fits' complexities limit the adjustment accuracy in this dimension, which is about 25 μm. The dimensions of the track on which the coils are assembled correlate to the magnetic trap's y position. The Y position alignment can be improved by moving the track to multiple different final positions, then nonresonantly ionizing Rb atoms in the trap. This method is shown in Figure 6.4(b). The cloud shape and the extraction slit in the back most trap electrode are combined represented by the Rb ion signal. The number of Rb+ ions arriving at the detector is optimized, indicating optimal alignment. The uncertainty in the Gaussian fit limits the alignment's uncertainty, which is about 50 micrometers. The track positioning system's accuracy is on the order of 10 μm, so alignment precision in this dimension is limited. By taking absorption images of the Rb trap whereas the electric trapping fields are active, can also be used for alignment in this dimension. Here, alignment is made easier by optimizing the number of atoms and the symmetry of the image. However, because of the previously mentioned issues with eddy currents in the dual-trap region, this latter method proved to be less precise (Patcharaphun & Mennig, 2005; Rezaei et al., 2009). The most difficult task is aligning the two traps along the Stark decelerator's longitudinal axis (z). In this case, the Rb lifetime in the dual trap is used to optimize the alignment, which is illustrated in figure 6. 4(c). Owing to this data, before taking lifetime measurements, the Rb trap is pre-exposed to the electrostatic trapping fields for 0.5 seconds. Which then fits a single exponential decay. This method reduces sensitivity to the rapid decay that occurs when electric fields [$\tau 1$ in Eq. (1)] are introduced and satisfy the collision-data protocol's requirements (described below). The Rb lifetime is maximized when optimally aligned because the Rb atoms are exposed to the smallest possible electric field strengths. Mechanical shimming of the track on the optical table is used to adjust the magnetic trap position along this axis. Within 50 μm, this position can be replicated. So, the absorption images in the dual-trap region, like those in the z dimension, can be used to verify approximate alignment.

6.6. EXTRACTING CROSS SECTIONS

As previously stated, the dual trap is a complex environment in which both species' population and spatial profiles change over time. It's impossible to build an analytic model that accurately represents ND3 density decay because there are so many dynamics going on at the same time. Instead, we use trajectory simulations combined with Monte Carlo techniques to model trapped ND3 decay and extract associated collision cross-sections, because this model integrates all demonstrated trap dynamics. Three simulations are required for each isotope combination to capture the full dynamics and accurately model the experiment.

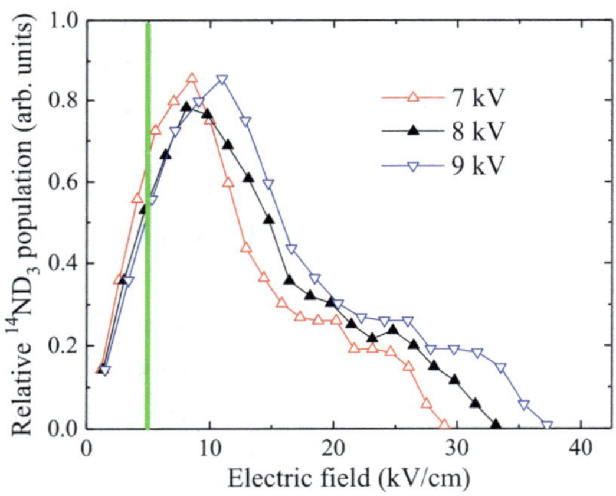

Figure 6.5. For three different trap voltage amplitudes, the average magnitude of the electric field sampled by molecules in the electrostatic trap was simulated. The vertically green line denotes the point at which the inelastic cross section's electric field enhancement is supposed to saturate.

Source: https://arxiv.org/abs/2001.02317#:~:text=8%20Jan%202020%5D-, Collisions%20Between%20Ultracold%20Atoms%20and%20Cold, a%20 Dual%20Electrostatic%2DMagnetic%20Trap&text=Simultaneous%20trapping%20offers%20significantly%20longer, sensitivity%20to%20rare%20collision%20events.

Before a final simulation, an initial simulation for each species is used to model the individual species' behavior in the dual trap.

REFERENCES

1. Ai, T., Wang, R., & Zhou, W., (2007). Effect of grafting alkoxysilane on the surface properties of Kevlar fiber. Polymer Composites, 28(3), 412–416.
2. Akay, M., & Barkley, D., (1991). Fiber orientation and mechanical behavior in reinforced thermoplastic injection moldings. Journal of Materials Science, 26(10), 2731–2742.
3. Andrews, G., & Peters, L., (1998). The psychometric properties of the composite international diagnostic interview. Social Psychiatry and Psychiatric Epidemiology, 33(2), 80–88.
4. Barbosa, S. E., & Kenny, J. M., (1999). Analysis of the relationship between processing conditions-fiber orientation-final properties in short fiber reinforced polypropylene. Journal of Reinforced Plastics and Composites, 18(5), 413–420.
5. Belgacem, M. N., Bataille, P., & Sapieha, S., (1994). Effect of corona modification on the mechanical properties of polypropylene/cellulose composites. Journal of Applied Polymer Science, 53(4), 379–385.
6. Briggs, D., Brewis, D. M., & Konieczo, M. B., (1976). X-ray photoelectron spectroscopy studies of polymer surfaces. Journal of Materials Science, 11(7), 1270–1277.
7. Chung, D. H., & Kwon, T. H., (2002). Fiber orientation in the processing of polymer composites. Korea-Australia Rheology Journal, 14(4), 175–188.
8. Chung, D. H., & Kwon, T. H., (2002). Numerical studies of fiber suspensions in an axisymmetric radial diverging flow: The effects of modeling and numerical assumptions. Journal of Non-Newtonian Fluid Mechanics, 107(1–3), 67–96.
9. Collyer, A. A., & Clegg, D. W., (1986). An introduction to fiber reinforced thermoplastics. In: Mechanical Properties of Reinforced Thermoplastics (Vol. 33, No. 1, pp. 1–8).
10. Fu, S. Y., Lauke, B., Mäder, E., Hu, X., & Yue, C. Y., (1999). Fracture resistance of short-glass-fiber-reinforced and short-carbon-fiber-reinforced polypropylene under Charpy impact load and its dependence on processing. Journal of Materials Processing Technology, 89, 501–507.
11. Gupta, M., & Wang, K. K., (1993). Fiber orientation and mechanical properties of short fiber reinforced injection molded composites: Simulated and experimental results. Polymer Composites, 14(5), 367–382.

12. Imihezri, S. S. S., Sapuan, S. M., Sulaiman, S., Hamdan, M. M., Zainuddin, E. S., Osman, M. R., & Rahman, M. Z. A., (2006). Mold flow and component design analysis of polymeric-based composite automotive clutch pedals. Journal of Materials Processing Technology, 171(3), 358–365.
13. Jagur-Grodzinski, J., (1992). Modification of polymers under heterogeneous conditions. Progress in Polymer Science, 17(3), 361–415.
14. Jeong, E., Lee, B. H., Doh, S. J., Park, I. J., & Lee, Y. S., (2012). Multifunctional surface modification of an aramid fabric via direct fluorination. Journal of Fluorine Chemistry, 141, 69–75.
15. Kadolph, S. J., & Casselman, K. D., (2004). In the bag: Contact natural dyes. Clothing and Textiles Research Journal, 22(1, 2), 15–21.
16. Kaplan, S. L., & Hansen, W. P., (1997). Gas plasma treatment of Kevlar [R] and spectra [R] fabrics for advanced composites. In: International Sample Technical Conference (Vol. 29, pp. 142–145).
17. Kharitonov, A. P., (2000). Practical applications of the direct fluorination of polymers. Journal of Fluorine Chemistry, 103(2), 123–127.
18. Labay, C., Canal, C., & García-Celma, M. J., (2010). Influence of corona plasma treatment on polypropylene and polyamide 6.6 on the release of a model drug. Plasma Chemistry and Plasma Processing, 30(6), 885–896.
19. Larsen, Å., (2000). Injection molding of short fiber reinforced thermoplastics in a center gated mold. Polymer Composites, 21(1), 51–64.
20. Leon, A. S., Franklin, B. A., Costa, F., Balady, G. J., Berra, K. A., Stewart, K. J., & Lauer, M. S., (2005). Cardiac rehabilitation and secondary prevention of coronary heart disease: An American heart association scientific statement from the council on clinical cardiology (subcommittee on exercise, cardiac rehabilitation, and prevention) and the council on nutrition, physical activity, and metabolism (subcommittee on physical activity), in collaboration with the American association of cardiovascular and pulmonary rehabilitation. Circulation, 111(3), 369–376.
21. Lin, T. K., Wu, S. J., Lai, J. G., & Shyu, S. S., (2000). The effect of chemical treatment on reinforcement/matrix interaction in Kevlar-fiber/bismaleimide composites. Composites Science and Technology, 60(9), 1873–1878.

22. Ma, C., Chen, D. L., Bhole, S. D., Boudreau, G., Lee, A., & Biro, E., (2008). Microstructure and fracture characteristics of spot-welded DP600 steel. Materials Science and Engineering: A, 485(1, 2), 334–346.
23. Malzahn, J. C., & Schultz, J. M., (1986). Transverse core fiber alignment in short-fiber injection-molding. Composites Science and Technology, 25(3), 187–192.
24. Margrave, J. L., & Lagow, R. J., (1979). Direct fluorination: A 'new' approach to fluorine chemistry. Prog. Inorg. Chem, 26, 161.
25. Mavrich, A. M., Ishida, H., & Koenig, J. L., (1995). Infrared mapping of surface-modified Kevlar fiber-reinforced epoxy systems. Applied Spectroscopy, 49(2), 149–155.
26. Mohanty, S., Nayak, S. K., Verma, S. K., & Tripathy, S. S., (2004). Effect of MAPP as a coupling agent on the performance of jute-PP composites. Journal of Reinforced Plastics and Composites, 23(6), 625–637.
27. Nguyen-Chung, T., & Mennig, G., (2001). Non-isothermal transient flow and molecular orientation during injection mold filling. Rheologica Acta, 40(1), 67–73.
28. Nguyen-Chung, T., Plichta, C., & Mennig, G., (1998). Flow disturbance in polymer melt behind an obstacle. Rheologica Acta, 37(3), 299–305.
29. Park, S. J., & Park, B. J., (1999). Electrochemically modified PAN carbon fibers and interfacial adhesion in epoxy-resin composites. Journal of Materials Science Letters, 18(1), 47–49.
30. Park, S. J., Seo, M. K., Ma, T. J., & Lee, D. R., (2002). Effect of chemical treatment of Kevlar fibers on mechanical interfacial properties of composites. Journal of Colloid and Interface Science, 252(1), 249–255.
31. Patcharaphun, S., & Mennig, G., (2005). Properties enhancement of short glass fiber reinforced thermoplastics via sandwich injection molding. Polymer Composites, 26(6), 823–831.
32. Patel, M., & Padhi, B. K., (1991). Formation of alumina whiskers using $Al_2(SO_4)_3$ and aluminum isopropoxide. Journal of Materials Science Letters, 10(21), 1243–1245.
33. Rezaei, F., Yunus, R., & Ibrahim, N. A., (2009). Effect of fiber length on thermomechanical properties of short carbon fiber reinforced polypropylene composites. Materials & Design, 30(2), 260–263.

34. Scola, D. A., & Brooks, C. S., (1970). Surface aspects of new fibers, boron, silicon carbide, and graphite. The Journal of Adhesion, 2(3), 213–237.
35. Senthilvelan, S., & Gnanamoorthy, R., (2006). Fiber reinforcement in injection-molded nylon 6/6 spur gears. Applied Composite Materials, 13(4), 237–248.
36. Singh, P., & Kamal, M. R., (1989). The effect of processing variables on microstructure of injection molded short fiber reinforced polypropylene composites. Polymer Composites, 10(5), 344–351.
37. Theocaris, P. S., & Papanicolaou, G. C., (1980). Variation of glass transition temperature with direction in unidirectional glass fiber-reinforced composites. Colloid and Polymer Science, 258(9), 1044–1051.
38. Wan, Z. H., Lin, J. Z., & You, Z. J., (2005). non-axisymmetric instability in the Taylor-Couette Flow of fiber suspension. Journal of Zhejiang University-Science A, 6(1), 1–7.
39. Wu, G. M., Hung, C. H., You, J. H., & Liu, S. J., (2004). Surface modification of reinforcement fibers for composites by acid treatments. Journal of Polymer Research, 11(1), 31–36.
40. Yadav, T., Brahmachari, G., Karmakar, I., Yadav, P., Agarwal, A., Mukherjee, V., & Dubey, N. K., (2021). Structural confirmation of biorelevant molecule N-iso-butyl, S-2-nitro-1-phenylethyl dithiocarbamate in gas phase and effect of fluorination. Chemical Physics Letters, 762, 138124.
41. Yue, C. Y., & Padmanabhan, K., (1999). Interfacial studies on surface-modified Kevlar fiber/epoxy matrix composites. Composites Part B: Engineering, 30(2), 205–217.
42. Zakir, M., Tsoi, J. K. H., Chu, C. H., Lung, C. Y. K., & Matinlinna, J. P., (2014). Bonding dissimilar materials in dentistry: A critical review. Reviews of Adhesion and Adhesives, 2(4), 413–432.
43. Zhu, Y., Ailane, N., Sala-Valdés, M., Haghighi-Rad, F., Billard, M., Nguyen, V., & Greco, C., (2017). Multi-factorial modulation of colorectal carcinoma cells motility-partial coordination by the tetraspanin Co-029/tspan8. Oncotarget, 8(16), 27454.

Chapter 7

Physics of Few-Body Single Ion in an Ultracold Bath

CONTENTS

7.1. INTRODUCTION

Cold chemistry emphasis the study of chemical reactions at temperatures 1 mK $\leq T \leq 1$ K (Doyle et al., 2004; Carr & Ye, 2009). At these temperatures, molecules, and atom's dynamics are probably dominated through pure quantum mechanical behavior, taking to captivating phenomena on chemical reactions like resonance effects. Additionally, the average kinetic energy of molecules and atoms is comparable with the usual energy shifts made by exterior fields. Therefore, it is probable to regulate the interaction and motion of molecules and atoms efficiently. Specifically, thanks to the progress of ion-neutral hybrid traps, one could study charged-neutral interactions in a well-regulated manner, paving the method to the discovery and analysis of new reaction mechanisms (Willitsch et al., 2008; Hall et al., 201; Dörfler et al., 2019).

Currently, ion-neutral hybrid traps function a more striving purpose (separately from cold chemistry): the study of impurity physics of a charged unit in a neutrals sea. Impurity physics is a traditional area in summarized matter physics, founded on the concept of quasiparticle: an impurity bath hybrid entity with specific dynamics. A key example is Landau and Pekar's work in which they noticed that charge carriers (impurity) interrelate with the vibrations of the bath (lattice), covering the charge carrier with the successive creation of a quasiparticle recognized as polaron (Tan'shyna, 2018; You et al., 2015). The polaron idea plays a vital role in understanding wonders of condensed matter physics like colossal magnetoresistance charge transport in organic semiconductors and high-temperature superconductivity (Ackers & Smith, 1985; Firsov, 2007). Certainly, it has been probable to study the polaron model after the model of material science thanks to molecular and atomic systems at ultracold temperatures (Kepple & Griem, 1968; Balewski et al., 2013).

Generally, the study of few-body procedures, like chemical reactions (a few-body procedure per sec), looked at as a topic sort out from condensed matter physics. Though, any many-body theory needs input from few-body procedures that could constrain, affect, and finally control a few of the system's many-body aspects. A key instance is a Rydberg impurity in an ultracold gas (Balewski et al., 2013; Karpiuk et al., 2015). In this situation, the Rydberg atom brings collective excitations in the ultracold bath throughout the Rydberg atom's lifetime. Though, it was perceived that the Rydberg lifetime is much littler when placed in a bath than in a vacuum. This was anonymous till it was realized that the quicker decay rate is a result of an inelastic few-body procedure and a reactive one (Schlagmüller et al., 2016; Chen et al., 2017). The inelastic procedure is known as l-changing collision, in which the Rydberg state finishes up in a high angular momentum state after striking with a neutral atom. The reactive procedure is the associative ionization (one of the chemi-ionization mechanisms in which the Rydberg electron is ionized, taking to the creation of a molecular ion (Mihajlov et al., 2012; Yamakita & Kai, 2019).

In ion-neutral hybrid traps, the extended range nature of the charged-neutral interface associated with neutral one might lead to novel and surprising many-body phenomena (Schurer et al., 2014; Astrakharchik et al., 2020). Though, a more dominant long-range interaction takes inevitably to chemical reactions, which disturbs the lifetime and state of the impurity (if it comprises internal degrees of freedom), as it has been lately shown (Schlagmüller et al., 2016; Mohammadi et al., 2021). Thus, to have a complete understanding of the many-body physics of a charged contamination in an ultracold bath, it is essential to characterize the inelastic and reactive procedures of the system, which, commonly, is further complicated through the existence of time-dependent trapping potentials.

7.2. A NECESSARY PREAMBLE

Before proceeding into this article's topic, we think it is necessary to emphasize that cold chemistry, as specified in the introduction, stands for any chemical reaction happening at temperatures between 1 K and 1 mK, where quantum mechanical impacts might play a substantial role on different sprinkling observables. Certainly, the significance of quantum mechanical impacts in a specified range of temperatures relies on the underlying inter-particle interaction. For example, let us undertake that the long-range interaction among two particles is given by $-C_n/R^n$, where C_n is the long-range coefficient and $n \geq 3$. The number of partial waves, l_{Scatt}, adding to any sprinkling observable at given collision energy, E_k, is given by:

$$l_{\text{Scatt}} = \left(\frac{2}{n-2}\right)^{\frac{n-2}{2n}} \sqrt{n\mu} C_n^{1/n} E_k^{\frac{n-2}{2n}}, \tag{1}$$

where μ is the decreased mass of the colliding partners. Thus, the inter-particle interface potential sets the number of partial waves appropriate for scattering observables at given collision energy.

Supposing that at l_{Scatt} ~ 20 quantum mechanical impacts are possibly averaged over, we have calculated the lowest collision energy at which a classical method is dependable (through Equation (1)) and the consequences for alkali-alkali ion collisions and alkali-alkali are presented in Figure 7.1. As a consequence, we observe that quantum mechanical effects govern cold chemistry processes including neutrals. Though, at the same system, ion-atom interactions in ion-neutral hybrid traps are well-suited situations for a classical trajectory method (Carr & Ye, 2009; Pérez-Ríos, 2019).

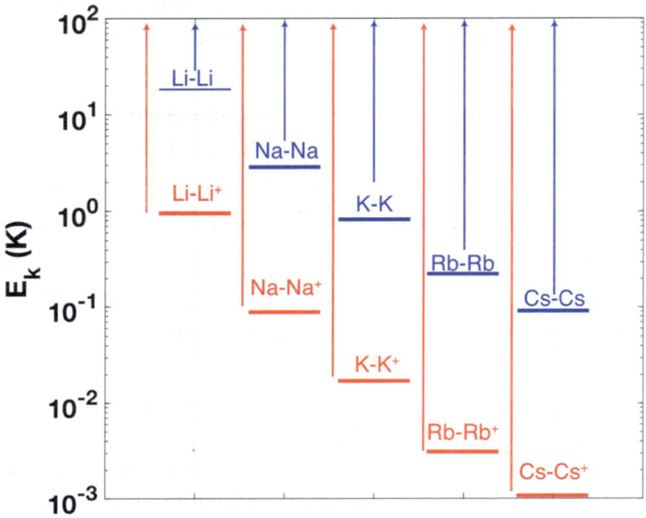

Figure 7.1. Lesser collision energy at which 20 partial waves add to the sprinkling observable. The lines beginning from the left of the markers mention ion-neutral interaction, however, the rest of the lines mention neutral-neutral interactions. The rising pointing arrows show the area of validity of a classical trajectory method for the sprinkling observables.

Source: https://journals.aps.org/pra/abstract/10.1103/PhysRevA.98.062707.

7.3. ANATOMIC ION IN A BATH OF ULTRACOLD ATOMS

A single ion stuck in a sea of ultracold atoms is the sample arena for the study of unusual impurity physics, and it has been widely studied primarily from a many-body perspective (Schurer et al., 2014; Astrakharchik et al., 2020). As a consequence, it is expected that a single ion in a bath of ultracold atoms might create a bound state with the nearby atoms leading to the creation of a 'mesoscopic molecular ion' (Schurer et al., 2014; Astrakharchik et al., 2020). Most of the literature regarding this topic emphasis on the system's ground state properties and dynamics, ignoring the motion of the ion, with the exemption of the work of Schurer et al. (2014), in which the motion of the ion in a harmonic trap is comprised.

An ion, A+, in interaction with a high-density gas of ultracold atoms practices three-body collisions with the atoms of the bath, B, leading to the creation of molecular ions through ion-atom-atom three-body recombination (TBR).

$A^+ + B + B \rightarrow AB^+ + B$,
or neutral molecules (2)

$$A^+ + B + B \rightarrow B_2 + A^+, \quad (3)$$

In three-body collisions comprising neutral and charged particles, the charged-neutral interaction exceeds the neutral-neutral interaction in the energetic merging to TBR, which explains two significant features. The initial is that molecular ions are specially made in comparison to neutral molecules as it is drawn on panel (i) of Figure 7.3 (Li et al., 2018; Pérez-Ríos & Greene, 2018). In other words, reaction channel (2) is overriding for ion-atom-atom collisions. The majority of molecular ions as the last product state has been experimentally established and partly sustained through complete quantum mechanical calculations about the creation of molecular anions in some systems (Krükow et al., 2016; Wang et al., 2017). Thus, it looks that our investigations at the cold regime might be extensible, under appropriate circumstances, to the ultracold realm. The second is that the ion-atom-atom TBR rate displays a threshold behavior as a purpose of the collision energy as (Li et al., 2018):

$$k_3(E_k) \propto E_k^{-3/4}, \quad (4)$$

where E_k means the collision energy. Equation (4) has been derived supposing a classical trajectory method in hyperspherical coordinates which is well suitable to study the dynamics of an atom-ion systems at cold temperatures (Pérez-Ríos et al., 2014). In specific, the motion of the nuclei is ruled through the electronic potential energy surface (PES) (upcoming from quantum chemistry calculations) by the solutions of Hamilton's equations. The legitimacy of Eq. (4) has been experimentally corroborated for Rb + Ba^+ + Rb collisions for $E_k \sim 10$ Mk (Krükow et al., 2016).

Furthermore, it has been probable to originate a first principle thermal-averaged TBR rate for A +A⁺+ A, where A is a noble gas atom, shown by (Li et al., 2018; Pérez-Ríos & Greene, 2018).

$$k_3(T) = \frac{1}{2(k_B T)^3}\int_0^\infty k_3(E_k)E_k^2\, e^{-\frac{E_k}{k_B T}}\, dE_k$$

$$= \frac{8\pi^2}{15}\frac{\Gamma(9/4)3^{1/4}}{\sqrt{2}(k_B T)^{3/4}}\frac{(2\alpha)^{5/4}}{\sqrt{m}}, \quad (5)$$

where m is the mass of the noble gas atom, α is the polarizability of the noble gas atom, T stands for the temperature of the gas, k_B is the Boltzmann constant, and (x) is the Euler gamma function of argument x. Equation (5) properly describes the reliance of the TBR rate as a role of the noble gas atom properties.

For three-body collisions, the collisional time could be projected as:

$$\tau = \frac{1}{\rho^2 k_3}, \quad (6)$$

where ρ is the density of the gas (in the situation of ion-atom-atom TBR is the density of atoms). For Rb +Ba^++ Rb crashes at cold temperature, the TBR rate is ~10–^{24} cm^6/s and supposing a usual atomic density of an atomic gas near to quantum degeneracy (ρ ~ 10^{14} cm^3), the collisional time is 100 μ s. In further words, after 100 μ s the ion impurity would grow into a molecular ion, therefore given a higher limit to the lifetime of the impurity and therefore limiting any many-body phenomena happening in the system (Krükow et al., 2016).

7.3.1. A Weakly Bound Molecular Ion in an Ultracold Bath of Atoms

The subsequent molecular ion later ion-atom-atom TBR seems in a feebly bound vibrational state whose binding energy relates with the collision energy (Krükow et al., 2016). The molecular ion might further strike with atoms of the bath; therefore, it could look like a novel class of impurity. The dynamics of this *novel* impurity could be studied from a quasi-classical trajectory (QCT) view in which Newton's law of classical mechanics explains the motion of the nuclei in the PES. In comparison, the early conditions for the trajectories are selected according to the colliding partners' quantum state by the Wentzel, Kramers, and Brillouin (WKB) or semi-classical approximation. The same approach is applied to map the last values of momentum and position of the trajectories in the phase-space into quantum states of the structure at hand (Carr & Ye, 2009; Pérez-Ríos, 2019).

A feebly bound molecular ion striking with an atom leads, separately from the expected elastic collisions to 3 probable outcomes. First, vibrational quenching arbitrated through an effective translational-vibrational energy transfer as $AB(v)^{+} + B \rightarrow AB(v')^{+}$ +B, with $v' \neq v$ where v appear for the vibrational state of the molecular ion as drawn in panel (ii) of Figure 7.3. Second, molecular creation by an exchange reaction as $AB^{+} + B \rightarrow B_2 + A^{+}$, as it is shown in panel (iii) of Figure 7.3 supposing that the charge is positioned at atom A and that the dynamics happen in a single PES. In further words, charge transfer reactions are not deliberated. Third, detachment of the molecular ion leading to 2 free atoms and 1 ion as $AB(v)^{+} + B \rightarrow A^{+} + B + B$, which is schematically presented in panel (iv) of Figure 7.3. From all these probable reaction channels, the most appropriate at cold temperatures is the vibrational quenching followed through the dissociation, which is only lived as long as the collision energy is greater than the molecular ion's binding energy. Though, the creation of neutral molecules is not related at cold temperatures.

As an instance, the vibrational quenching cross-section calculated through a QCT formalism for RbBa+(v)+ Rb as a role of the collision energy for diverse vibrational states is revealed in Figure 7.2. In this figure, it is seen that the vibrational quenching cross-section is generally free of the vibrational state of the molecular ion, and its tendency approves with the predictions founded on the Langevin capture model, $\sigma_L(E_k) = \pi\sqrt{2\alpha/E_k}$ which is signified as the black dashed line.

The regions where the QCT outcomes deviate from the Langevin prediction relate to collision energies greater than the molecular ion's binding energy, represented as the vertical dashed lines in Figure 7.2. Certainly, the agreement between the Langevin prediction and vibrational quenching cross-section translates into a very effective vibrational-translational energy, which could be rationalized in terms of the adiabaticity limit (Carr & Ye, 2009; Pérez-Ríos, 2019).

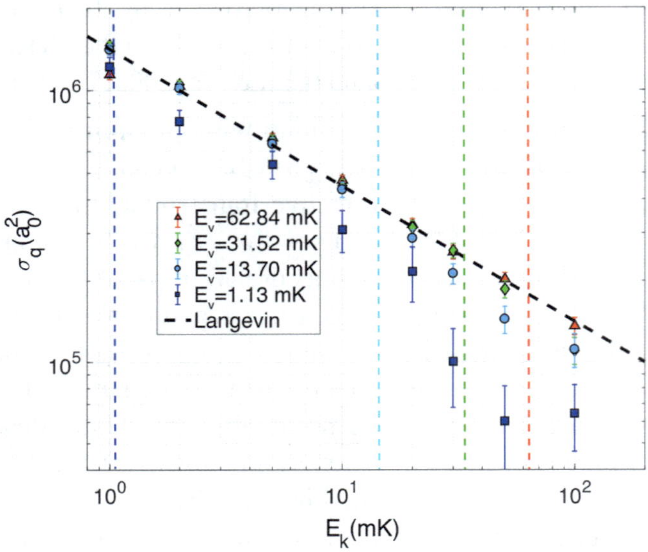

Figure 7.2. Quenching cross-section for the crash BaRb+(v)+Rb → BaRb Rb as a role of the collision energy (Ek). The diverse binding energies (Ev) of the vibrational states v are signified by diverse symbols as indicated on the legend. The dashed line denotes the Langevin cross-section. The vertical dashed lines appear for the binding energy of the earlier vibrational states of the molecular ion.

Source: https://journals.aps.org/pra/abstract/10.1103/PhysRevA.98.062707.

7.3.2. The Role of External Laser Sources

Each single ion-neutral hybrid trap trial needs diverse laser sources to laser-cool ions and to capture, hold, and manipulate ultracold atoms. These laser sources are normally supposed to have a little effect on the dynamics of a single ion in a bath of ultracold atoms. Though, it has been revealed that the trapping lasers of a magneto-optical trap (MOT) having ultracold atoms might lead to an improvement of the rate of charge-transfer reactions (Hall et al., 2011). Thus, laser sources might play an appropriate role in the study of a charged impurity in a bath of ultracold atoms.

The main instance of this phenomenology is a single ion engrossed in a high-density ultracold atomic gas (Mohammadi et al., 2021). In this situation, as explained above, a feebly bound molecular ion emerges because of an ion-atom-atom.

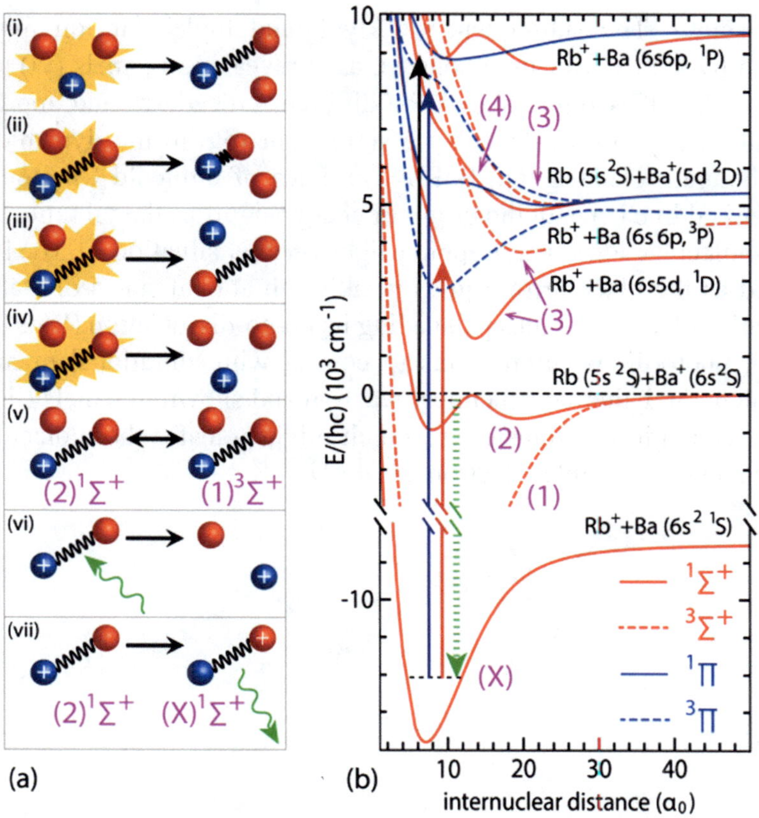

Figure 7.3. Collision channels (non-radiative and radiative) of a feebly bound molecular ion in a high-density ultracold gas.

Source: https://journals.aps.org/prresearch/abstract/10.1103/PhysRevResearch.3.013196.

Note: Panel (i) signifies ion-atom-atom three-body recombination; panel (ii) shows for a vibrational quenching of a molecular ion, panel (iii) is an alternate chemical reaction in which a molecule seems like the product state; panel (iv) signifies a dissociation procedure, panel (v) is a spin-flip collision, panel (vi) is related with a photo-dissociation procedure; and panel (vii) shows for a radiative decay procedure. The right panel displays the potential energy curves appropriate for the different radiative-assisted procedures in BaRb+. Solid line displays the possible photodetachment transition for1064 nm light, however, the broken solid lines display possible photo-dissociation transitions for 493 and 650 nm light. The dashed arrow shows radiative relaxation to the electronic ground state.

TBR reaction. The subsequent weakly bound molecular ion might be detached by the exterior light sources, as is revealed in panels (vi) and (b) of Figure 7.3. This figure shows the different procedures and appropriate potential energy curves that play a significant role in the dynamics of a feebly bound molecular ion, BaRb+, in a sea of ultracold Rb atoms (the black dashed line beneath the origin of energy signifies the vibrational state of the molecular ion). The trapping light (in this situation of 1064 nm is signified as the black arrow) joins weakly vibrational states of BaRb+ to dissociative electronic states persuading its photo-dissociation (Ríos, 2020). Though, photo-dissociation wants to contest with vibrational quenching, detachment, and alternate chemical reaction and shown in panels (ii-iv), as well as with spin-flip transitions, which might transfer the molecular ion into the metastable triplet electronic state$(1)^3\Sigma^+$.

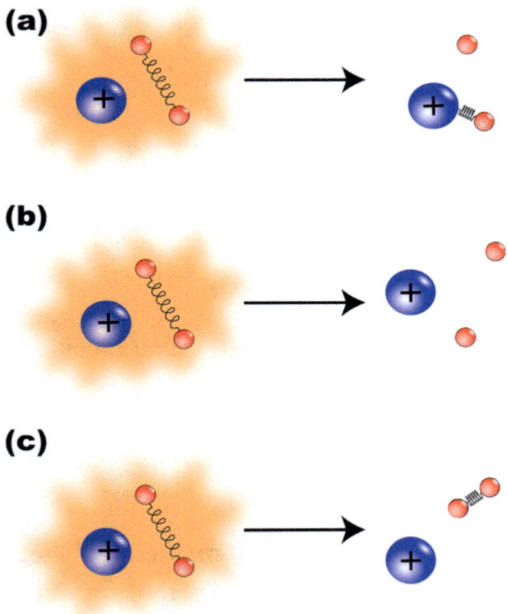

Figure 7.4. A diagram of the dynamics of an atomic ion striking with a molecule. Panel (a) stands for the molecular ion creation process, panel (b) for the detachment of the molecule, and panel (c) represents the vibrational quenching of the molecule.

Source: https://www.tandfonline.com/doi/full/10.1080/00268976.2021.188163 7.

Furthermore, the feebly bound molecular ion might decay into an extremely vibrational state of the $(X)^{1}+$ ground electronic state through spontaneous emission. Lastly, the molecular ion could relax into intensely bound vibrational states cheers to radiative decay processes or vibrational quenching collisions, and lastly, it could be photo-dissociated through the lasers employed to laser-cool Ba+. The role of each of the non-radiative and radiative procedures has been studied utilizing QCT results for the vibrational quenching, detachment, and alternate reactions, a semi-classical method for the spin-flip transitions, and a full-quantum action of the photo-dissociation cross-section (Mohammadi et al., 2021). This complicated cold reaction network is experimentally validated by looking at the significance of diverse ionic species as a function of time (Mohammadi et al., 2021). Lastly, we would like to focus out that the role of exterior laser sources on few-body procedures has also been noticed in molecular ion-atom systems (Dörfler et al., 2019).

REFERENCES

1. Ackers, G. K., & Smith, F. R., (1985). Effects of site-specific amino acid modification on protein interactions and biological function. Annual Review of Biochemistry, 54(1), 597–629.
2. Astrakharchik, G. E., Ardila, L. A., Schmidt, R., Jachymski, K., & Negretti, A., (2020). Ionic polaron in a Bose-Einstein condensate. arXiv preprint arXiv:2005.12033.
3. Balewski, J. B., Krupp, A. T., Gaj, A., Peter, D., Büchler, H. P., Löw, R., & Pfau, T., (2013). Coupling a single electron to a Bose-Einstein condensate. Nature, 502(7473), 664–667.
4. Carr, L. D., & Ye, J., (2009). Special Issue of New J. Phys., 11, 055009.
5. Chen, Z., Wang, Y., Li, T., Tian, L., Qiu, Y., Inomata, K., & You, J. Q., (2017). Single-photon-driven high-order sideband transitions in an ultra strongly coupled circuit-quantum-electrodynamics system. Physical Review A, 96(1), 012325.
6. Dörfler, A. D., Eberle, P., Koner, D., Tomza, M., Meuwly, M., & Willitsch, S., (2019). Long-range versus short-range effects in cold molecular ion-neutral collisions. Nature Communications, 10(1), 1–10.
7. Doyle, J., Friedrich, B., Krems, R. V., & Masnou-Seeuws, F., (2004). Special issue on cold molecules with an editorial: Quo Vadis cold molecules. Eur Phys. J. D, 31, 149.
8. Firsov, Y. A., (2007). Polarons in advanced materials. Small Polarons: Transport Phenomena, 63, 1–16.
9. Hall, F. H., Aymar, M., Bouloufa-Maafa, N., Dulieu, O., & Willitsch, S., (2011). Light-assisted ion-neutral reactive processes in the cold regime: Radiative molecule formation versus charge exchange. Physical Review Letters, 107(24), 243202.
10. Hickman, A. P., (1978). Theory of angular momentum mixing in Rydberg-atom-rare-gas collisions. Physical Review A, 18(4), 1339.
11. Hirzler, H., & Pérez-Ríos, J., (2021). Rydberg atom-ion collisions in cold environments. Physical Review A, 103(4), 043323.
12. Hulea, I. N., Fratini, S., Xie, H., Mulder, C. L., Iossad, N. N., Rastelli, G., & Morpurgo, A. F., (2006). Tunable Fröhlich polarons in organic single-crystal transistors. Nature Materials, 5(12), 982–986.
13. Karpiuk, T., Brewczyk, M., Rzążewski, K., Gaj, A., Balewski, J. B., Krupp, A. T., & Pfau, T., (2015). Imaging single Rydberg electrons in a Bose-Einstein condensate. New Journal of Physics, 17(5), 053046.

14. Kepple, P., & Griem, H. R., (1968). Improved Stark profile calculations for the hydrogen lines Hα, Hβ, Hγ, and Hδ. Physical Review, 173(1), 317.
15. Krükow, A., Mohammadi, A., Härter, A., Denschlag, J. H., Pérez-Ríos, J., & Greene, C. H., (2016). Energy scaling of cold atom-atom-ion three-body recombination. Physical Review Letters, 116(19), 193201.
16. Lammers, J., Weimer, H., & Hammerer, K., (2016). Open-system many-body dynamics through interferometric measurements and feedback. Physical Review A, 94(5), 052120.
17. Li, J., Liu, J., Luo, L., & Gao, B., (2018). Three-body recombination near a narrow Feshbach resonance in Li 6. Physical Review Letters, 120(19), 193402.
18. Mannella, N., Rosenhahn, A., Booth, C. H., Marchesini, S., Mun, B. S., Yang, S. H., ... & Fadley, C. S. (2004). Direct observation of high-temperature polaronic behavior in colossal magnetoresistive manganites. Physical review letters, 92(16), 166401.
19. Mihajlov, A. A., Srećković, V. A., Ignjatović, L. M., & Klyucharev, A. N., (2012). The chemi-ionization processes in slow collisions of Rydberg atoms with ground-state atoms: Mechanism and applications. Journal of Cluster Science, 23(1), 47–75.
20. Mohammadi, A., Krükow, A., Mahdian, A., Deiß, M., Pérez-Ríos, J., Da Silva, Jr. H., & Denschlag, J. H., (2021). Life and death of a cold $BaRb^+$ molecule inside an ultracold cloud of Rb atoms. Physical Review Research, 3(1), 013196.
21. Pérez-Ríos, J., & Greene, C. H., (2018). Universal temperature dependence of the ion-neutral-neutral three-body recombination rate. Physical Review A, 98(6), 062707.
22. Pérez-Ríos, J., & Robicheaux, F., (2016). Rotational relaxation of molecular ions in a buffer gas. Physical Review A, 94(3), 032709.
23. Pérez-Ríos, J., (2019). Vibrational quenching and reactive processes of weakly bound molecular ions colliding with atoms at cold temperatures. Physical Review A, 99(2), 022707.
24. Pérez-Ríos, J., Ragole, S., Wang, J., & Greene, C. H., (2014). Comparison of classical and quantal calculations of helium three-body recombination. The Journal of Chemical Physics, 140(4), 044307.
25. Podzorov, V., Menard, E., Borissov, A., & Kiryukhin, V., (2004). In: Rogers, J. A., & Gershenson, M. E., (eds.), Phys. Rev. Lett. (Vol. 93, p. 86602).

26. Ríos, J. P., (2020). Ultracold Rydberg atoms and ultralong-range Rydberg molecules. In: An Introduction to Cold and Ultracold Chemistry (pp. 137–153). Springer, Cham.
27. Schlagmüller, M., Liebisch, T. C., Engel, F., Kleinbach, K. S., Böttcher, F., Hermann, U., & Greene, C. H., (2016). Ultracold chemical reactions of a single Rydberg atom in a dense gas. Physical Review X, 6(3), 031020.
28. Schurer, J. M., Schmelcher, P., & Negretti, A., (2014). Ground-state properties of ultracold trapped bosons with an immersed ionic impurity. Physical Review A, 90(3), 033601.
29. Stoecklin, T., Halvick, P., Gannouni, M. A., Hochlaf, M., Kotochigova, S., & Hudson, E. R., (2016). Explanation of efficient quenching of molecular ion vibrational motion by ultracold atoms. Nature Communications, 7(1), 1–8.
30. Tan'shyna, A., (2018). Lev landau. Ukraine, Kharkiv, uipt. World, 3, 46.
31. Wang, B. B., Han, Y. C., Gao, W., & Cong, S. L., (2017). Cold atom-atom-ion three-body recombination of 4 He–4 He–X−(X= H or D). Physical Chemistry Chemical Physics, 19(34), 22926–22933.
32. Wang, J., Liu, X. J., & Hu, H., (2019). Roton-induced Bose polaron in the presence of synthetic spin-orbit coupling. Physical Review Letters, 123(21), 213401.
33. Willitsch, S., Bell, M. T., Gingell, A. D., & Softley, T. P., (2008). Chemical applications of laser-and sympathetically-cooled ions in ion traps. Physical Chemistry Chemical Physics, 10(48), 7200–7210.
34. Yamakita, Y., & Kai, N., (2019). Classical trajectory calculations for state-resolved penning ionization reactions of polycyclic aromatic hydrocarbon C10H8 in collision with He*(23S). Molecular Physics, 117(21), 3184–3193.
35. Yan, Z. Z., Ni, Y., Robens, C., & Zwierlein, M. W., (2020). Bose polarons near quantum criticality. Science, 368(6487), 190–194.
36. You, Z. Q., Mewes, J. M., Dreuw, A., & Herbert, J. M., (2015). Comparison of the Marcus and Pekar partitions in the context of nonequilibrium, polarizable-continuum solvation models. The Journal of Chemical Physics, 143(20), 204104.

Chapter 8

Quantum Transport in Ultracold Atoms

CONTENTS

8.1. INTRODUCTION

As it is a very fast and appropriate way to carry and manipulate energy and information thus, the transport of electrons has played an important role in modern technology. Quantum mechanics describes, through band theory, in nature the amazing variability of electrical resistivity, ranging up to $10^{16}\,\Omega$m in insulators and up to $10^{-8}\,\Omega$m in metals, an important fact for applications. Thus this is an exciting effect via itself since, superficially, through which we came to know that in transport, the Coulomb interaction plays a relatively little role, a fact produced inside the outline of the Landau-Fermi liquid theory, which indicates that the charge carriers form a Fermi sea of feasible interacting quasiparticles (Ji et al., 2006). It is rather clear that both the Coulomb interaction and the quantum nature of the carriers or other kinds of interaction can have exciting and potentially valuable consequences (Huang et al., 2007).

Take an example of superconductivity. In common superconductors, the electron-phonon interaction increases the dissipationless electronic transport. While common superconductors are reliably defined through the Bardeen-Cooper-Schrieffer (BCS) theory which indirectly depends on the Landau-Fermi liquid model, the most exciting superconductors along with high critical temperature (high-T_c superconductors) display strong signals of a collapse of the quasiparticle picture and a quantitative theory thereof is still deficient (Bardeen et al., 1957; Reid et al., 2016). The fractional quantum Hall effect is another situation in which Coulomb connections are evident most dramatically, and guide to a strong example of a breakdown of the Landau-Fermi liquid picture (Zhang, 1992), in which a two-dimensional electron gas acts as a gas of quasiparticles along with slight charge and only at the edges the transport of currents happens. The quantum Hall effect and superconductivity indicate how transport is certainly one of the strongest probes at our removal to examine the matter in its several phases. They also indicate how our present knowledge of the properties of matter, in specific its transport properties, is deeply founded on models where the connections are treated approximately and the parameters of validity of such estimates are often unclear.

A new route to manipulate transport and probe of the quantum matter was identified when rarefied atomic gases (with typical density $n \sim 10^{13}$–10^{15} cm^{-3} compared to $n \sim 10^{22}$cm^{-3} in solid-state systems and very cold (down to a few nano kelvins) was employed to comprehend the condensate of bosons predicted in 1924 by Einstein and Bose (Davis et al., 1995; Isoard & Pavloff,

2020). These ultracold gases are particularly controllable and clean: as an example, the interparticle boundary takes a very simple form, specifically, that of an isotropic contact interface parametrized just through the two-body scattering length, that can be visibly tuned thus retrieving the whole range from the noninteracting boundary to the greatly interacting one (Isoard & Pavloff, 2020). On Bose-Einstein condensates, elegant researches permit complete analyzes of interrelating bosons both out of equilibrium and in the ground state. The contract with theory is quantitatively outstanding, showing that it is one of the few examples of a many-body quantum system where the effect of connections is well understood. Soon after, the even more stimulating mission of cooling Fermi gases below the degeneracy temperature was achieved and noninteracting Fermi gases were understood (DeMarco & Jin, 1999; Truscott et al., 2001). Thus, noninteracting fermions and bosons are no more ideal concepts and ultracold atoms become experimentally available.

These accomplishments marked the start of a novel field of ultracold atoms and research—as these very cold and rarefied atomic gases are called—are so flexible and controllable that they can be measured as "quantum simulators" specifically physical systems that can copycat the performance of other quantum states of matter whose understanding confronts conventional means (Cirac & Zoller, 2012).

8.2. COMPARISON BETWEEN ULTRACOLD GASES AND SOLID STATE SYSTEMS

In common electronics by comparing transport phenomena in ultracold atomic gases to similar ones, their similarities and differences, brief in Table 8.1, should be considered. Afterward, when several related topics are discussed, their importance will be appreciated. The most important is that, because of the incarceration and extremely low density and temperature when electronic systems are coupled with the external environment, ultracold gases are essentially very well isolated from it shown in Figure 8.1. Initially, an ultracold gas could be driven through small external perturbations at rest out of equilibrium, frequently quite far from it. Based on linear response theory, this is a problem for common methods to nonequilibrium dynamics but simultaneously an opportunity for studying transport in commands outside linear response that was formerly distant (Galperin et al., 2006; Hofferberth et al., 2007). Regardless, cold-atom systems still permit to test of common linear-response theory (Hazlett et al., 2013). The formation of highly-excited metastable states along with "negative temperature" is an extreme situation compared with the motional degrees of freedom (DOF) an outcome that reignited the discussion on the steady thermodynamic descriptions of isolated systems and the concept of negative temperatures (Chien & Ventra, 2012; Dunkel & Hilbert, 2014).

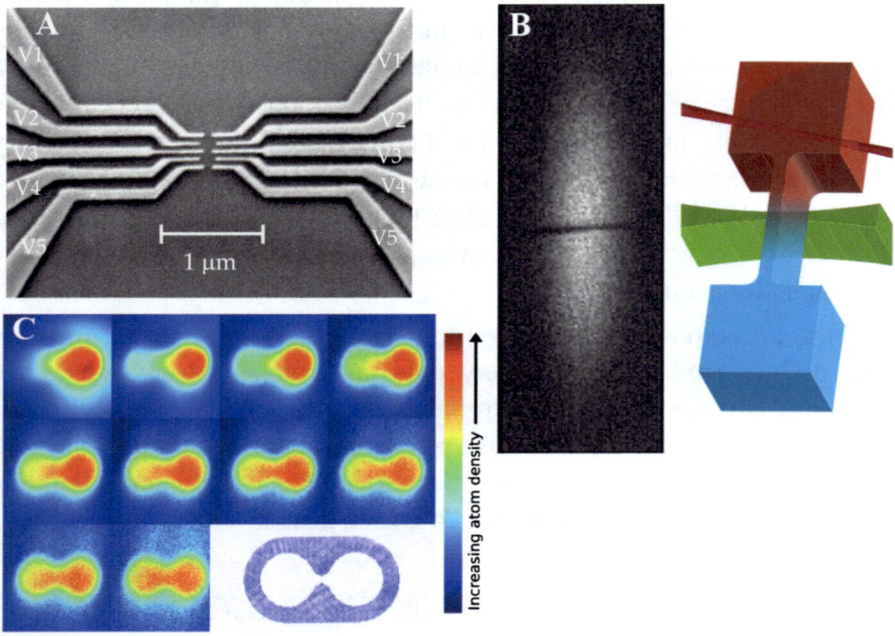

Figure 8.1. In solid-state, the contrast of transport experiments (A) and cold-atom (B, C) systems.

Source: https://www.nature.com/articles/nphys3531.

Note: In (A), The quantum dot molecule is demonstrated and is well-defined through gating leads and is linked to the bulk of the two-dimensional electron gas. Because of the coupling among the bulk and dots the electron gas, this is an open system. As compared, ultracold atomic systems are generally closed systems since particles only move around inside the system. In (B), one may be taken as the two sides of the intersection as reservoirs, nevertheless, the complete particle number is preserved. In (C), the atoms flow from one side to the other in the ultracold analog of an RC-circuit applied by constraining the atoms in the dumbbell-shaped region shown in the lower right (from Lee et al., 2013).

Table 8.1. Summary of the Similarities and Differences Among Cold Atoms and Conventional Electronic Systems. Here Np is the Particle Number and N is the Total Number of Lattice Sites

	Electronics	**Cold atoms**
Confinement	Positively-charged ionic lattice	Optical or magnetic potentials, usually harmonic
Temperature	10^{-6}–10^{2}K	10^{-9}–10^{-6}K
Particle statistics	Fermionic (electrons)	Bosonic or fermionic
Density	$\sim 10^{22}$ cm^{-3}	10^{13}–10^{15} cm^{-3}
Number of components	Two (spin-1/2)	Variable, depending on atom species
Conventional approach	Top-down	Bottom-up
Natural setup	Open	Isolated from the environment
Thermodynamic limit	N_p/N = constant, $N \rightarrow \infty$.	N and N_p can be tuned separately.
Lattice geometry	Controlled by chemical composition, pressure, etc.	Highly tunable
Tunneling time scale	$\sim 10^{-15}$ s or less	$\sim 10^{-3}$ s
Interaction	Coulomb interactions	Tunable. The noninteracting limit can be accessed.
Driving force	Electromagnetic field, temperature gradient, etc.	artificial gauge fields Gravity, magnetic field gradients, local heating,
Disorder	Generally unavoidable	Almost none but can be engineered
Main observable	Charge current	Density profile

A state in which the other observables or average current remain persistent inside a finite period is known as a quasi-steady-state which is generally an important hypothesis in any theory of transport (Huang et al., 2007; Chern et al., 2014). Nevertheless, away from the linear response pattern, it is not clear that in an isolated system, a quasi-steady state always establishes and what cases are needed. For fermions, this is the common behavior, but in the case of other systems, complicated dynamics or oscillatory behavior can arise rather than a quasi-steady-state (Chien et al., 2014). In nanoscopic and mesoscopic systems the Landauer theory of transport employed to model transport like the one in Figure 8.1 (a) it assumes both the presence of a steady-state and that the energy reservoirs and large particle associated to the conduction channel are in the thermal equilibrium state explained by the Fermi-Dirac distribution (Huang et al., 2007).

As we can see in Figure 8.1 (b) and (c), in ultracold atom systems, the "reservoirs" contains a much lesser size, and general expectations in common theories of transport may not essentially hold (Huang et al., 2007). The particle distribution generally develops along with the time when a current is flowing, in an isolated finite system, and a motionless Fermi-Dirac distribution might not be accurate, particularly if the dynamics are determined in a particle-number preserving fashion as in ultracold gases (Strohmaier et al., 2007; Sommer et al., 2011). In Figure 8.2 (b), an example of the dynamical evolution of particle distributions in the reservoirs is shown in which an exciting thought is that the quasi-steady state current is still flowing in a similar direction while the initial particle number imbalance can reverse. A "microcanonical" method to transport which is not dependent on the artificial distinction among the system of interest and environment, in cold atoms it seems to be better suited to examine transport and has already as a theoretical tool proved its worth (Huang et al., 2007; Chern et al., 2014). Certainly, the microcanonical method acts as a link between the conventional transport setup and isolated systems; in the conventional method, the diverse load and power source are present, while in an isolated system, the particle number is preserved exactly. It is suitable for the study of a quantum quench, specifically an unexpected change of a factor employed to excite a response, that is the very common sort of experiment made with ultracold gases (Schneider et al., 2012).

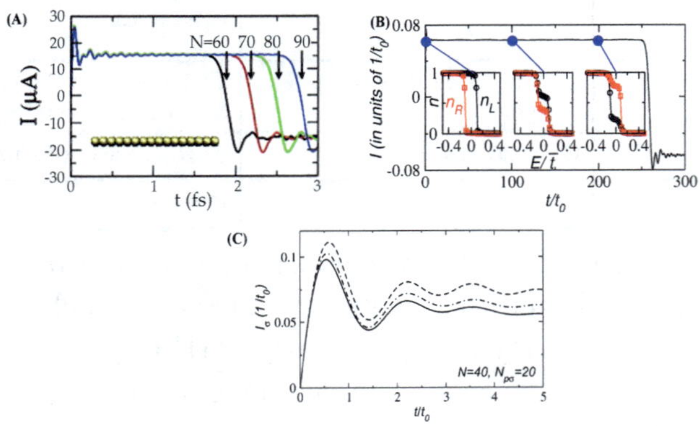

Figure 8.2. (a) A QSSC (quasi-steady-state current) can arise in an isolated quantum system of noninteracting fermions. (b) In an isolated system, even when a quasi-steady state current is present, the particle distributions nR and nL of the two "finite reservoirs" evolve in time. (c) In the occurrence of onsite interactions, a quasi-steady state current still survives.

Source: https://iopscience.iop.org/article/10.1088/1367–2630/15/6/063026/meta.

By employing common linear response theory, the transport phenomena that are now analyzed may give more information when fully microcanonical analyzes are applied (Brantut et al., 2012; Krinner et al., 2015). Instead, cold-atoms systems in which atoms in a trap are constantly loaded and detached with a terminator beam, such as the cold-atom analog of a battery, have been perceived. In this situation, suitable alterations to the microcanonical formalism should be applied.

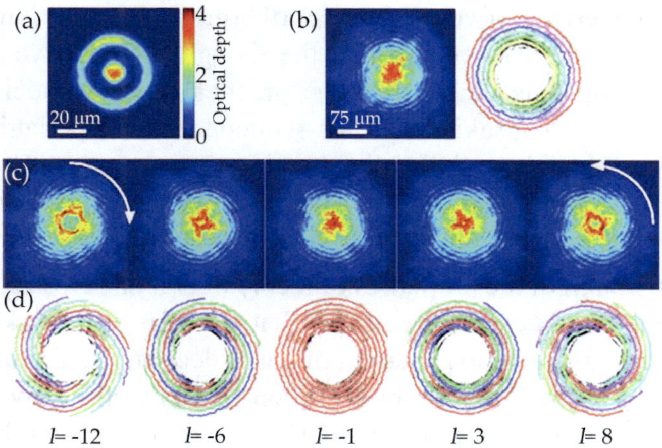

Figure 8.3. (a) Absorption imaging of a ring-shaped condensate along with additional condensate in the middle. The interfering observed upon releasing the inner condensate can be employed to measure the total current flowing in the ring. (b) displays the interfering pattern attained in the absence of current, whereas the spiraling patterns shown in(c) and (d) agree to the finite amount of angular momentum quanta l shown at the bottom.

Source: https://journals.aps.org/prx/abstract/10.1103/PhysRevX.4.031052.

8.3. TRANSPORT IN WEAKLY-INTERACTING GASES

In solid-state systems, the transport phenomena could be reproduced successfully through a large number of experiments with ultracold gases, generally inside an outline of weakly-interacting or non-interacting particles (Hazlett et al., 2013). In solid-state systems, these phenomena can be occasionally challenging if not unfeasible to access experimentally, and ultracold gases are an outstanding opportunity to return to them, simultaneously putting the foundations for undertaking more cultured models and benchmarking available experimental methods for the measurement, manipulation, and preparation the of these systems. Few important examples are Bloch oscillations which in solid-state systems resist a clear demonstration but have been smartly realized using cold atoms in an optical lattice, Landau-Zener tunneling, and Wannier-Stark ladder in periodic potentials (Morsch & Oberthaler, 2006). Because of the absence of a substantial amount of disorder, these illustrations have been probable, an intrinsic characteristic of cold atoms. Furthermore, the addition of a finite collision rate to the carriers permits the thought of negative differential conductance, which is reckoned to be practically well-modeled through the classical Tsu and Esaki model for semiconductor superlattices (Rode, 1975). In systems with spatially reliant on interactions, negative differential conductance has been projected to occur also, which can be understood in ultracold gases.

To the research of the properties of feebly interacting Bose gases a large body of work has been dedicated, for that the Gross-Pitaevskii equation has occurred as a simple explanation of gases deep in the shortened phase, specifically for Bose-Einstein condensation (BEC) well below the critical temperature (Isoard & Pavloff, 2020). This equation explains the agitation of the wave function employed by a macroscopic number of particles with the connections accounted for through a nonlinear term. These systems are controllable both experimentally and theoretically and are clear appearances of the wave nature of matter. Outstanding reviews of the subject are offered, and thus we will not debate the topic further (Isoard & Pavloff, 2020). For transport-related researchers of ultracold gases, exciting applications of this method are projected to yield a direct comparison with those in solid-state systems, whereas the general harmonic incarceration of cold atoms presents inhomogeneity that obscures the theory and the understanding of data (Lin et al., 2009).

In order to study the superfluid properties, the ultracold gases are also an outstanding laboratory. Both fermionic and bosonic superfluids are accessible and the series from weak to strong connections can be accessed (Isoard & Pavloff, 2020). It is coming to know that a smoking gun of superfluid transport is the quantization of vorticity and actually in atomic Bose-Einstein condensates direct imaging of arrays of vortex cores have been detected and in ultracold gases provided a definite proof of superfluidity. In a ring geometry, a tenacious current is an equilibrium property which has been found in superfluid helium and superconductors through employing a magnetic flux via the rotation and by the ring (Deaver & Fairbank, 1961).

Developments in determining the laser-induced potential have made it likely to produce atomic determined currents in single-component bosons (as shown in Figure 8.3) along with in spinor bosons and to perceive in real time the phase-slip procedures that guide to the decline of the current. According to theoretical analyzes, the current is predictable to endure in a wide range of inhomogeneity and interactions. Hysteresis of the quantized tenacious currents has been described as an outcome that has been vague in both the much-studied superconductors and superfluid helium. Furthermore, an optical potential blockade introduced in the ring act as a Josephson junction, whose current-phase interaction can be measured, such as through intrusive the ring-shaped condensate with additional condensate positioned in the middle of the ring (as shown in Figure 8.3). Introducing two of these junctions/barriers permits the understanding of the ultracold gas equivalent to a SQUID, a device with stimulating applications such as a high-precision rotation sensor. The superfluid fraction, which should be differentiated from the condensate portion, of ultracold atomic gases can be taken into account via the means of the second sound or by employing an equivalent of the classical Andronikashvili experiment.

Via the methods to understand artificial spin-orbit coupling fields and artificial gauge, the toolbox of cold-atom study has been lately improved which in ultracold gases enhance an entirely new dimension to the study of transport (Lin et al., 2009). By employing the berry phase encouraged by dressing atomic states along with laser fields, the artificial gauge fields are obtained, thus avoiding the restriction of charge neutrality of cold atoms. On a Bose-Einstein condensate, the consequence of an artificial magnetic field has been just taken into account and interpreted as a form of superfluid Hall effect. By employing a constant nonabelian gauge field, that is equal to a spin-orbit connection term, the bosonic equivalent to the spin Hall effect has been noticed. In lattices, numerous methods have also been employed to present artificial magnetic fluxes in the form of the typical Peierls replacement employing either lattice shaking, Raman-assisted hopping, and synthetic dimensions (Atala et al., 2014).

In the research works explained in Raman-assisted bounding has been employed to understand the Harper Hamiltonian involving in the mixture of a great uniform (pseudo-) magnetic flux along with a lattice potential and characterized through the fractal performance of the spectrum as a function of the magnetic field. In these research works, the fractal spectrum was not detected. Nonetheless, the dynamics of the atoms can be envisaged in real-time and is along with the existence of a large (pseudo-)magnetic field. The artificial magnetic field is another exciting characteristic of Raman-assisted bounding that can have differing signs for different spin components, which shows that time-reversal invariance is conserved, an important factor for a cold-atom execution of the newly revealed quantum spin Hall effect.

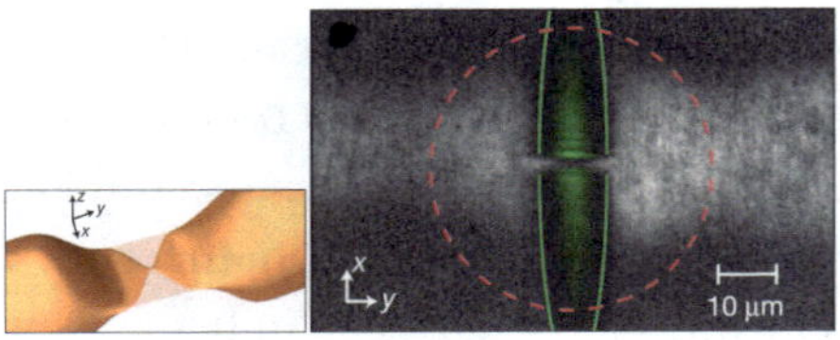

Figure 8.4. (Right) False-color image of the atomic density (green). (Left) A schematic 3-D view of the atomic quantum point contact employed to measure the phenomenon of quantized conductance.

Source: https://www.nature.com/articles/nature14049.

Note: Left: For fermionic ^{6}Li atoms, A laser in the TEM_{01} mode is employed to yield a thin conducting layer in an elongated atomic trap. A tightening laterally the x-direction is lithographically printed through another laser beam. The resultant channel supports insufficient transverse modes that can be selectively populated.

Right: The intensity profile of the laser beam is being indicated by the red dashed line employed to tune the chemical potential and therefore the transverse mode population in the channel.

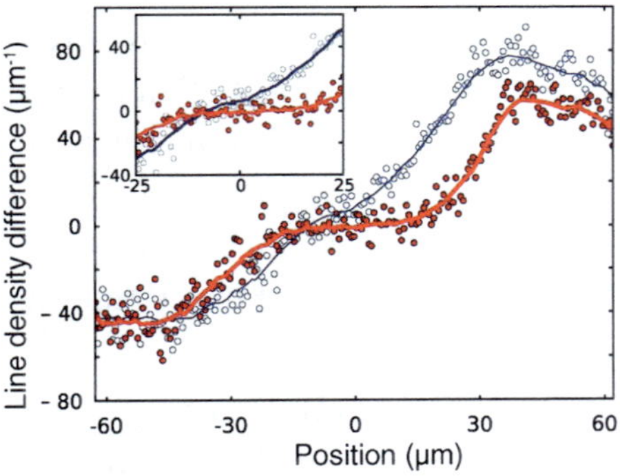

Figure 8.5. The variance of the density profiles among a current-carrying state and a state with zero current in the region of the channel contraction in the setup is shown in Figure 8.4.

Source: https://science.sciencemag.org/content/337/6098/1069.abstract.

Note: of the empty blue dots, the disorder has been artificially introduced, whereas in the case the red full dots refer to a ballistic channel.

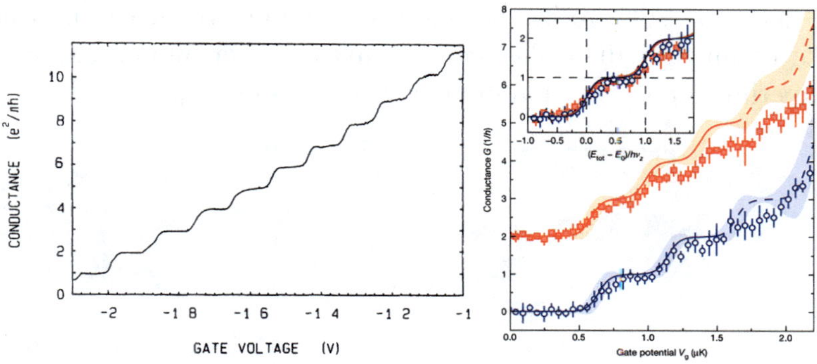

Figure 8.6. In a 2D dimensional electron Quantized conductance plateaus gas [2DEG] (left), and in the cold atom setup shown in Figure 8.4 (right).

Source: https://www.nature.com/articles/nature14049.

Note: In units of $G_0 = 2e^2/h$, the electronic transport the conductance is restrained whereas the equal unit in the case of the neutral matter is $G_{0,\ neut.}$ =1/h. Related parameters in the 2DEG: temperature T =0.6K, contraction width in correspondence of the first plateau W ≈ 20 nm, areal density n $=3.56\times10^{11}$ cm^{-2}, mean free path (l) = 8.5 μm, Fermi wavelength $\lambda_F \approx 40$ nm. Appropriate parameters in the cold atom experiment: Fermi wavelength λ_F ≈2μm, constriction width W ≈1.5 μm, T =42 nK, collision mean free path (l)=12 mm, density n =1013 cm^{-3}.

8.4. TRANSPORT IN STRONGLY-INTERACTING GASES

By applying external magnetic fields, the tunability of the interactions between atoms and in artificial periodic potentials the capability to trap an atomic gas are two significant ingredients for retrieving sturdily interacting states of quantum gases (Isoard & Pavloff, 2020). Via tuning the effective contact interface, the particular Fermi gas, in which two-body bound states initiate to emerge, has been analyzed and realized in great detail. Essential transport coefficients like the shear viscosity of these greatly interrelating systems have been taken into account. At low temperatures, the shear viscosity η is projected to be on the order of times the density whereas the scaling $\eta \sim T^{3/2}$ is expected at higher temperatures, and both outcomes have been verified. The ratio η/s, in which the entropy density *s* can be resultant from the equation of state, is near enough an estimated universal lower bound η/s ~ ~/k_B. There has been a discussion on whether in η or η/s the singular Fermi gas will display a minimum as its temperature reduces, and a new experiment proposes a minimum in η/s but not in η.

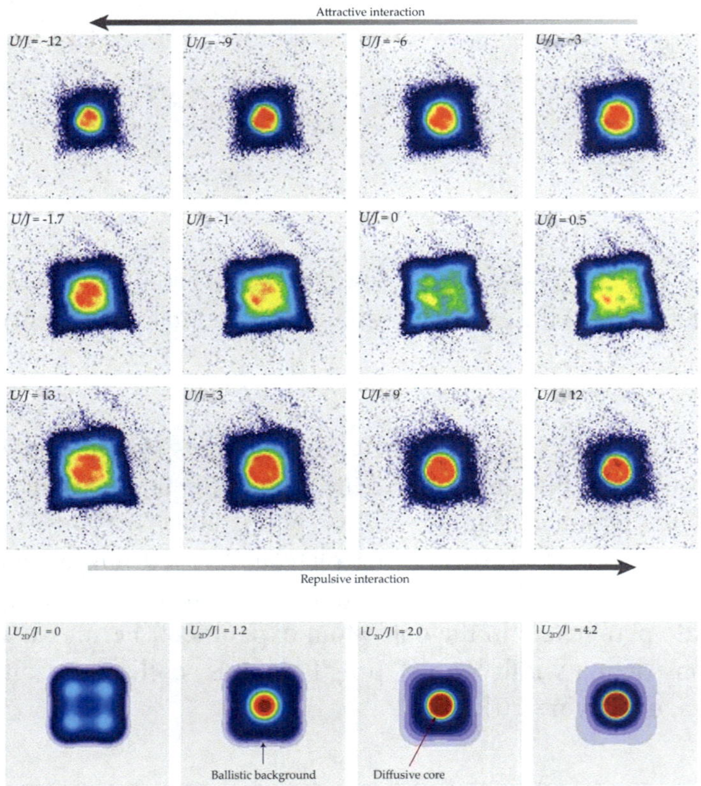

Figure 8.7. Free expansion of fermionic atoms in a two-dimensional optical lattice for various values of the Hubbard interaction U/J from attractive to repulsive.

Source: https://www.nature.com/articles/nphys2205.

Note: In the noninteracting case, the gas expands ballistically and the shape changes from spherically symmetric to square, which is the symmetry of the underlying lattice. Instead, in the interacting case, a large portion of the gas remains spherically symmetric, an indication of relaxation to thermal equilibrium, while only the low-density tails have a square shape due to the ballistic motion. An approach based on the Boltzmann equation can qualitatively capture the dynamics (bottom).

REFERENCES

1. Atala, M., Aidelsburger, M., Lohse, M., Barreiro, J. T., Paredes, B., & Bloch, I., (2014). Observation of chiral currents with ultracold atoms in bosonic ladders. Nature Physics, 10(8), 588–593.
2. Bakr, W. S., Gillen, J. I., Peng, A., Fölling, S., & Greiner, M., (2009). A quantum gas microscope for detecting single atoms in a Hubbard-regime optical lattice. Nature, 462(7269), 74–77.
3. Bardeen, J., Cooper, L. N., & Schrieffer, J. R., (1957). Theory of superconductivity. Physical Review, 108(5), 1175.
4. Bettles, R. J., Gardiner, S. A., & Adams, C. S., (2015). Cooperative ordering in lattices of interacting two-level dipoles. Physical Review A, 92(6), 063822.
5. Brantut, J. P., Meineke, J., Stadler, D., Krinner, S., & Esslinger, T., (2012). Conduction of ultracold fermions through a mesoscopic channel. Science, 337(6098), 1069–1071.
6. Braun, S., Ronzheimer, J. P., Schreiber, M., Hodgman, S. S., Rom, T., Bloch, I., & Schneider, U., (2013). Negative absolute temperature for motional degrees of freedom. Science, 339(6115), 52–55.
7. Bushong, N., Sai, N., & Di Ventra, M., (2005). Approach to steady-state transport in nanoscale conductors. Nano Letters, 5(12), 2569–2572.
8. Chang, G., & Roth, C. B., (2001). Structure of MsbA from E. coli: A homolog of the multidrug resistance ATP binding cassette (ABC) transporters. Science, 293(5536), 1793–1800.
9. Cheneau, M., Barmettler, P., Poletti, D., Endres, M., Schauß, P., Fukuhara, T., & Kuhr, S., (2012). Light-cone-like spreading of correlations in a quantum many-body system. Nature, 481(7382), 484–487.
10. Chern, G. W., Chien, C. C., & Di Ventra, M., (2014). Dynamically generated flat-band phases in optical Kagome lattices. Physical Review A, 90(1), 013609.
11. Chien, C. C., & Di Ventra, M., (2012). Dynamical crossover between the infinite-volume and empty-lattice limits of ultra-cold fermions in 1D optical lattices. EPL (Europhysics Letters), 99(4), 40003.
12. Chien, C. C., Di Ventra, M., & Zwolak, M., (2014). Landauer, Kubo, and microcanonical approaches to quantum transport and noise: A comparison and implications for cold-atom dynamics. Physical Review A, 90(2), 023624.

13. Chien, C. C., Zwolak, M., & Di Ventra, M., (2012). Bosonic and fermionic transport phenomena of ultracold atoms in one-dimensional optical lattices. Physical Review A, 85(4), 041601.
14. Cirac, J. I., & Zoller, P., (2012). Goals and opportunities in quantum simulation. Nature Physics, 8(4), 264–266.
15. Dalibard, J., Gerbier, F., Juzeliūnas, G., & Öhberg, P., (2011). Colloquium: Artificial gauge potentials for neutral atoms. Reviews of Modern Physics, 83(4), 1523.
16. Davis, K. B., Mewes, M. O., Andrews, M. R., Van, D. N. J., Durfee, D. S., Kurn, D. M., & Ketterle, W., (1995). Bose-Einstein condensation in a gas of sodium atoms. Physical Review Letters, 75(22), 3969.
17. Deaver, Jr. B. S., & Fairbank, W. M., (1961). Experimental evidence for quantized flux in superconducting cylinders. Physical Review Letters, 7(2), 43.
18. DeMarco, B., & Jin, D. S., (1999). Onset of fermi degeneracy in a trapped atomic gas. Science, 285(5434), 1703–1706.
19. Dunkel, J., & Hilbert, S., (2014). Consistent thermostatistics forbids negative absolute temperatures. Nature Physics, 10(1), 67–72.
20. Galperin, M., Nitzan, A., & Ratner, M. A., (2006). Resonant inelastic tunneling in molecular junctions. Physical Review B, 73(4), 045314.
21. Goldman, N., Dalibard, J., Dauphin, A., Gerbier, F., Lewenstein, M., Zoller, P., & Spielman, I. B., (2013). Direct imaging of topological edge states in cold-atom systems. Proceedings of the National Academy of Sciences, 110(17), 6736–6741.
22. Gotlibovych, I., Schmidutz, T. F., Gaunt, A. L., Navon, N., Smith, R. P., & Hadzibabic, Z., (2014). Observing properties of an interacting homogeneous Bose-Einstein condensate: Heisenberg-limited momentum spread, interaction energy, and free-expansion dynamics. Physical Review A, 89(6), 061604.
23. Hazlett, E. L., Ha, L. C., & Chin, C., (2013). Anomalous Thermoelectric Transport in Two-dimensional Bose Gas. arXiv preprint arXiv:1306.4018.
24. Hofferberth, S., Lesanovsky, I., Fischer, B., Schumm, T., & Schmiedmayer, J., (2007). Nonequilibrium coherence dynamics in one-dimensional Bose gases. Nature, 449(7160), 324–327.
25. Huang, Z., Chen, F., D'agosta, R., Bennett, P. A., Di Ventra, M., & Tao, N., (2007). Local ionic and electron heating in single-molecule junctions. Nature Nanotechnology, 2(11), 698.

26. Isoard, M., & Pavloff, N., (2020). Departing from thermality of analogue Hawking radiation in a Bose-Einstein condensate. Physical Review Letters, 124(6), 060401.
27. Ji, A., Li, C., & Cao, Z., (2006). Ternary Cu 3 N Pd x exhibiting invariant electrical resistivity over 200 K. Applied Physics Letters, 89(25), 252120.
28. Krinner, S., Stadler, D., Husmann, D., Brantut, J. P., & Esslinger, T., (2015). Observation of quantized conductance in neutral matter. Nature, 517(7532), 64–67.
29. Lee, J. G., McIlvain, B. J., Lobb, C. J., & Hill, III. W. T., (2013). Analogs of basic electronic circuit elements in a free-space atom chip. Scientific Reports, 3(1), 1–4.
30. Lin, Y. J., Compton, R. L., Jiménez-García, K., Porto, J. V., & Spielman, I. B., (2009). Synthetic magnetic fields for ultracold neutral atoms. Nature, 462(7273), 628–632.
31. Mandel, O., Greiner, M., Widera, A., Rom, T., Hänsch, T. W., & Bloch, I., (2003). Coherent transport of neutral atoms in spin-dependent optical lattice potentials. Physical Review Letters, 91(1), 010407.
32. McKay, D. C., Meldgin, C., Chen, D., & DeMarco, B., (2013). Slow thermalization between a lattice and free Bose gas. Physical Review Letters, 111(6), 063002.
33. Morsch, O., & Oberthaler, M., (2006). Dynamics of Bose-Einstein condensates in optical lattices. Reviews of Modern Physics, 78(1), 179.
34. Paredes, B., Widera, A., Murg, V., Mandel, O., Fölling, S., Cirac, I., & Bloch, I., (2004). Tonks-Girardeau gas of ultracold atoms in an optical lattice. Nature, 429(6989), 277–281.
35. Peotta, S., Chien, C. C., & Di Ventra, M., (2014). Phase-induced transport in atomic gases: From superfluid to mott insulator. Physical Review A, 90(5), 053615.
36. Reid, J. P., Tanatar, M. A., Luo, X. G., Shakeripour, H., De Cotret, S. R., Juneau-Fecteau, A., & Taillefer, L., (2016). Doping evolution of the superconducting gap structure in the underdoped iron arsenide $Ba_{1-x}K_xFe_2As_2$ revealed by thermal conductivity. Physical Review B, 93(21), 214519.
37. Rode, D. L., (1975). Low-field electron transport. In: Semiconductors and Semimetals (Vol. 10, pp. 1–89). Elsevier.

38. Ronzheimer, J. P., Schreiber, M., Braun, S., Hodgman, S. S., Langer, S., McCulloch, I. P., & Schneider, U., (2013). Expansion dynamics of interacting bosons inhomogeneous lattices in one and two dimensions. Physical Review Letters, 110(20), 205301.
39. Salger, T., Kling, S., Hecking, T., Geckeler, C., Morales-Molina, L., & Weitz, M., (2009). Directed transport of atoms in a Hamiltonian quantum ratchet. Science, 326(5957), 1241–1243.
40. Sato, M., Takahashi, Y., & Fujimoto, S., (2009). Non-Abelian topological order in s-wave superfluids of ultracold fermionic atoms. Physical Review Letters, 103(2), 020401.
41. Schmidutz, T. F., Gotlibovych, I., Gaunt, A. L., Smith, R. P., Navon, N., & Hadzibabic, Z., (2014). Quantum Joule-Thomson effect in a saturated homogeneous Bose gas. Physical Review Letters, 112(4), 040403.
42. Schneider, U., Hackermüller, L., Ronzheimer, J. P., Will, S., Braun, S., Best, T., & Rosch, A., (2012). Fermionic transport and out-of-equilibrium dynamics in a homogeneous Hubbard model with ultracold atoms. Nature Physics, 8(3), 213–218.
43. Soltan-Panahi, P., Struck, J., Hauke, P., Bick, A., Plenkers, W., Meineke, G., & Sengstock, K., (2011). Multi-component quantum gases in spin-dependent hexagonal lattices. Nature Physics, 7(5), 434–440.
44. Sommer, A., Ku, M., Roati, G., & Zwierlein, M. W., (2011). Universal spin transport in a strongly interacting fermi gas. Nature, 472(7342), 201–204.
45. Strohmaier, N., Takasu, Y., Günter, K., Jördens, R., Köhl, M., Moritz, H., & Esslinger, T., (2007). Interaction-controlled transport of an ultracold fermi gas. Physical Review Letters, 99(22), 220601.
46. Sun, K., Liu, W. V., Hemmerich, A., & Sarma, S. D., (2012). Topological semimetal in a fermionic optical lattice. Nature Physics, 8(1), 67–70.
47. Tarruell, L., Greif, D., Uehlinger, T., Jotzu, G., & Esslinger, T., (2012). Creating, moving and merging Dirac points with a fermi gas in a tunable honeycomb lattice. Nature, 483(7389), 302–305.
48. Truscott, A. G., Strecker, K. E., McAlexander, W. I., Partridge, G. B., & Hulet, R. G., (2001). Observation of fermi pressure in a gas of trapped atoms. Science, 291(5513), 2570–2572.
49. Vidmar, L., Langer, S., McCulloch, I. P., Schneider, U., Schollwöck, U., & Heidrich-Meisner, F., (2013). Sudden expansion of mott insulators in one dimension. Physical Review B, 88(23), 235117.

50. Zhang, S. C., (1992). The Chern-Simons-landau-Ginzburg theory of the fractional quantum hall effect. International Journal of Modern Physics B, 6(01), 25–58.
51. Zozulya, A. A., & Anderson, D. Z., (2013). Principles of an atomtronic battery. Physical Review A, 88(4), 043641.

Index

Q

R

S

T

U

Printed in the United States
by Baker & Taylor Publisher Services